Confronting Hunger in the USA

Food insecurity in the US is a critical issue that is experienced by approximately 15% of the population each year. Hunger is not caused by an inability to produce enough food for the population, but is instead a manifestation of federal agricultural policies that support the overproduction of commodity crops and neoliberal social policies that seek to lower the amount of benefits dispersed to those in need. This book focuses on how four different food-based community programs address both the physical sensation of hunger as well as the political and economic disempowerment that work against the ability of people experiencing food insecurity to mobilize as a political force.

Confronting Hunger in the USA argues that most food programs do more to create community among their volunteers than among program participants and tend to reinforce neoliberal understandings of citizenship. Community food programs reach out to the most vulnerable members of society in caring and gentle ways and often use the language of alternative economies to articulate a different relationship between the individual and the state. However, the projects in this study act as individual pieces of the state's insufficient social safety net and are only beginning to articulate a new relationship between food and society.

Adam M. Pine is Assistant Professor in the Department of Geography, Urban, Environment and Sustainability Studies at the University of Minnesota Duluth, USA.

Critical Food Studies
Series editor: Michael K. Goodman
University of Reading, UK

The study of food has seldom been more pressing or prescient. From the intensifying globalization of food, a world-wide food crisis and the continuing inequalities of its production and consumption, to food's exploding media presence, and its growing re-connections to places and people through 'alternative food movements', this series promotes critical explorations of contemporary food cultures and politics. Building on previous but disparate scholarship, its overall aims are to develop innovative and theoretical lenses and empirical material in order to contribute to – but also begin to more fully delineate – the confines and confluences of an agenda of critical food research and writing.

Of particular concern are original theoretical and empirical treatments of the materialisations of food politics, meanings and representations, the shifting political economies and ecologies of food production and consumption and the growing transgressions between alternative and corporatist food networks.

For a full list of titles in this series, please visit
https://www.routledge.com/Critical-Food-Studies/book-series/CFS

Careful Eating: Bodies, Food and Care
Edited by Emma-Jayne Abbots, Anna Lavis and Luci Attala

Food Pedagogies
Edited by Rick Flowers and Elaine Swan

Food and Media
Practices, Distinctions and Heterotopias
Edited by Jonatan Leer and Karen Klitgaard Povlsen

Confronting Hunger in the USA
Searching for Community Empowerment and Food Security in Food Access Programs
Adam M. Pine

Confronting Hunger in the USA

Searching for Community Empowerment and Food Security in Food Access Programs

Adam M. Pine

LONDON AND NEW YORK

First published 2017
by Routledge

2 Park Square, Milton Park, Abingdon, Oxfordshire OX14 4RN
52 Vanderbilt Avenue, New York, NY 10017

Routledge is an imprint of the Taylor & Francis Group, an informa business

First issued in paperback 2020

Copyright © 2017 Adam M. Pine

The right of Adam M. Pine to be identified as author of this work has been asserted by him in accordance with sections 77 and 78 of the Copyright, Designs and Patents Act 1988.

All rights reserved. No part of this book may be reprinted or reproduced or utilised in any form or by any electronic, mechanical, or other means, now known or hereafter invented, including photocopying and recording, or in any information storage or retrieval system, without permission in writing from the publishers.

Notice:
Product or corporate names may be trademarks or registered trademarks, and are used only for identification and explanation without intent to infringe.

British Library Cataloguing-in-Publication Data
A catalogue record for this book is available from the British Library

Library of Congress Cataloging-in-Publication Data
Names: Pine, Adam, 1975- author.
Title: Confronting hunger in the USA : searching for community empowerment and food security in food access programs / Adam M. Pine.
Description: Abingdon, Oxon ; New York, NY : Routledge, 2016. |
Includes bibliographical references and index.
Identifiers: LCCN 2016001729| ISBN 9781472411198 (hardback) |
ISBN 9781472411204 (e-book)
Subjects: LCSH: Food relief--United States. | Hunger--United States. |
Food security--United States--Citizen participation. | Community-based social services--United States.
Classification: LCC HV696.F6 P56 2016 | DDC 363.8/830973--dc23LC
record available at http://lccn.loc.gov/2016001729

ISBN: 978-1-4724-1119-8 (hbk)
ISBN: 978-0-367-66826-6 (pbk)

Typeset in Times New Roman
by Saxon Graphics Ltd, Derby

Contents

	List of Illustrations	vi
	List of Abbreviations	vii
	Acknowledgments	ix
1	The Struggle to Build Community and Feed Society	1
2	Food Security, the Industrial Food System, and Community in the US	28
3	Creating Community and Empowerment in Community-based Food Programs	48
4	The Emerging Alternative Economies in Community-based Food Programs	80
5	Neoliberalism and the Porous Continuum of Care for the Food Insecure	111
6	Looking for Paths to Food Access and Solidarity	135
	Bibliography	154
	Index	167

Illustrations

Figures

1.1 City of Duluth 5
1.2 Population of Duluth, 1870–2010 6

Tables

1.1 Food Distinctions between Ruby's Food Pantry, SHARE, CHUM, and Seeds of Success 12
1.2 Labor Distinctions between Ruby's Food Pantry, SHARE, CHUM, and Seeds of Success 13
1.3 Participant Experience Distinctions between Ruby's Food Pantry, SHARE, CHUM, and Seeds of Success 13
1.4 Ideological Distinctions between Ruby's Food Pantry, SHARE, CHUM, and Seeds of Success 15
2.1 Main Federal Food Support Programs. *Source:* Food Research Action Center, State of the States Report, 2008 34
3.1 Selected Participant Demographic Data from CHUM, Ruby's Food Pantry, and SHARE 52
3.2 Cooking by Participants in CHUM, Ruby's Food Pantry, and SHARE 66
4.1 Alternative Provisioning Strategies used by Participants in CHUM, Ruby's Food Pantry, and SHARE 103
4.2 Other Services that Participants in CHUM, Ruby's Food Pantry, and SHARE Would Like to See Offered 107
5.1 Characteristics of Food-insecure Participants in CHUM, Ruby's Food Pantry, and SHARE 126

Abbreviations

AAA	Agricultural Adjustment Administration
ABAWD	Able-bodied Adults Without Dependents
AFDC	Aid for Dependent Children
CAD	Community Action Duluth
CBPR	Community-based Participatory Research
CCA	Culture Centered Approach
CDBG	Community Development Block Grant
CHUM	Churches United in Ministry
CSA	Community Supported Agriculture
DYES	Duluth Youth Employment Service
EBT	Electronic Benefit Transfer
EITC	Earned Income Tax Credit
EWG	Environmental Working Group
FARA	Food Access Research Atlas
FFALP	Fair Food Access Lincoln Park
FNB	Food Not Bombs
FNS	Food and Nutrition Service
FSRC	Federal Surplus Relief Corporation
FTS	Farm to School
GFB	Good Food Box
GLSLS	Great Lakes St. Lawrence Seaway
JRLC	Joint Religious Legislative Coalition
LISC	Local Initiatives Support Corporation
MFAP	Minnesota Food Assistance Program
MNSure	Minnesota Health Insurance Marketplace
NAPS	Nutritional Access Program for Seniors
NGO	Non-governmental Organization
RFP	Ruby's Food Pantry
PRWORA	Personal Responsibility and Work Opportunity Reconciliation Act
SHARE	Self-help and Resource Exchange
SNAP	Supplemental Nutrition Assistance Program
SoS	Seeds of Success
SSI	Supplemental Security Income

TANF	Temporary Assistance for Needy Families
TEFAP	The Emergency Food Assistance Program
TRoTR	Te Runanga o Te Rarawa
USAID	United States Agency for International Development
USDA	United States Department of Agriculture
WIC	The Special Supplemental Nutrition Program for Women, Infants, and Children

Acknowledgments

This work could not have been completed without the generous support of the Institute for Advanced Study at the University of Minnesota Twin Cities that funded this research through the Abundance and Scarcity University Symposium (2010–2012) as a collaborative interdisciplinary research project. Special thanks to Ann Waltner and Susannah Smith. In addition, this research was funded by the Center for Community and Regional Research (CCRR) at the University of Minnesota Duluth.

Special thanks to all of the volunteers, participants, and coordinators at Seeds of Success, SHARE, Ruby's Food Pantry, and CHUM who were so generous with their time over the course of this research. Each of these organizations operates with skeletal staff and collectively improve the food security of thousands of residents of Duluth.

Thanks are also in order for the colleagues and students at the University of Minnesota Duluth who helped with this research. In the Department of Geography, Urban and Sustainability Studies Kate Carlson, Nathan Clough, Pat Farrell, Randel Hanson and the Sustainable Agriculture Project, Linda Klint, Mike Mageau, and Stacey Stark provided support for this work. Elsewhere in the University John Bennett provided immeasurable support. In addition, the students in the Geography Senior Seminar provided a sounding board for this research and Canyon Bachan, Kathleen Hammer, Charles Cochrane, and Aliina Charging Hawk helped to conduct interviews, enter data, and take care of the logistics of this research.

My biggest debt of thanks goes to Rebecca de Souza—my research partner and constant source of joy—without whom this project would just be a page of random notes about co-ops, SNAP benefits, and the problem of surplus agriculture production. I am blessed to have you in my life.

1 The Struggle to Build Community and Feed Society

> *"Food is just such a nice comfort thing, if you had a grandmother she would say 'eat, eat.'"*
>
> Tracy, CHUM Food Shelf volunteer

> *"No, our system here is good. I have nothing to say about Duluth and the way they help people here. Even the homeless, they always say 'There is no reason why you should be hungry in Duluth because you have all these places that feed.' All you got to do is show up, and you're not going to get turned down."*
>
> Alan, CHUM Food Shelf participant

> *"I don't have any money to go buy groceries. So I will either have to stop taking medication and buy food, or stop buying food and buy medication. With Ruby's Pantry I don't need to make that decision, I don't have to, I can just come here and have something to eat and be able to pay for my medications."*
>
> Heidi, Ruby's Food Pantry participant

Providing food to the hungry is a simple project that millions of people in the US take part in every day. From food shelves to soup kitchens and from food buying clubs to community gardens, civil society efforts to forestall hunger are viewed as an unquestioned good: they ensure that those who are poor still have enough food to not go to bed hungry and provide an outlet for those with the time or ability to give back to their community. But providing food for the marginalized is more complex than simply ladling soup into a bowl: it demands the creation of an organization, the drafting of a mission statement, the creation of a budget to oversee the project, the recruitment of volunteers, and an analysis of existing state and civil society welfare programs to determine how these new efforts will relate to the existing social safety net. Jesus teaches that "it is more blessed to give than to receive" and his followers and others in secular society have followed this injunction spiritedly opening up charities that provide nourishment to local and international victims of hunger. How do we understand civil society responses to hunger and what are the prospects of these institutions challenging the fundamental inequities in our food system that create hunger?

2   The Struggle to Build Community

Fifteen years ago Janet Poppendieck in *Sweet Charity* (1999) outlined a withering critique of the largest of these anti-hunger institutions: the food shelf. In her work she detailed how the food shelf system emerged in the 1980s as a civil society response to deep Reagan-era cuts in the social safety net as churches and civic groups responded to fill the need the very poor had for food. This was one of the first manifestations of the "shadow state" (Wolch 1990) as charitable groups and organizations in the voluntary sector stepped in to plug a gaping hole left by the downsizing state. Poppendieck argued that the US produces more than enough food to adequately nourish the population and the food shelf parcels out just enough of this overabundance so that clients escape persistent hunger, but do not have a stable food supply. Worse still, the managers of food shelves are potential anti-poverty activists, but they are so busy stocking shelves and soliciting donations they do not have time to advocate for social policies that will really transform the lives of their clients, and the more institutionalized charitable organizations become, the more the state can safely cede this aspect of the social safety net to others. Finally, the food shelf provided a place for volunteers to spend time together and give back to their community while food shelf clients were treated in a highly stigmatized way and provided with no sense of community or empowerment. As she summed up her argument:

> We are becoming attached to our charitable food programs and increasingly unable to envision a society that wouldn't need them. We are so busy building bigger, better programs to deliver food to the hungry, and to raise the funds and other resources necessary to continue and expand our efforts in response to the rising need, that we are losing sight of both the underlying problem and possible solutions.
>
> (1999, 17)

The food shelf system provided agribusiness a way to dispose of their surplus production, a place for volunteers to spend time together, but all the food insecure got was a small sack of unhealthy food.

Today Poppendieck's critique of the food shelf system has only grown more profound and complicated as the shadow state has grown larger and more professionalized in response to government cutbacks in poor relief. Feeding America, "the nation's leading domestic hunger-relief organization," partners with local food pantries to deliver food to households experiencing food insecurity and has grown into a multibillion-dollar operation that provides food to an astounding 46 million people each year (Feeding America 2013). It is now a permanent component of the shadow state with branches in all 50 states and a strong media and political presence. Further, as economic inequality has increased (Reich 2012) and the state has continued to reduce welfare benefits to the poor under the twin mantels of neoliberalism and workfarism (Peck 2001), the food shelf is just one part of a complex and entrenched system of civil society, voluntary sector, and government programs that work together to provide nourishment to the 14% of the US population that experience food insecurity each year

(Coleman-Jensen et al. 2014). On the civil society side soup kitchens, food shelves, food buying clubs and food salvage programs provide nourishment for low-income people and provide spaces where people who are concerned about the welfare of others can put their ideals into action. Collectively this community of activists is known as the anti-hunger community and their focus has traditionally been on addressing the short-term needs that individuals have for food, and not on fundamentally changing the long-term position of the food insecure (Allen 1999; Hall 2006). In addition, the government program Supplemental Nutritional Access Program (SNAP, formerly known as food stamps) is used by 34 million people in the US and costs the federal government about 50 billion dollars a year, while other feeding programs—such as the Special Supplemental Nutrition Program for Women, Infants, and Children (WIC)—serves millions more. These institutions work together to provision—albeit poorly—the millions of Americans that experience food insecurity each year.

Broadly, this book explores the problematic way in which food access programs in the US are organized: they fail to create food security or empower participants to push for their own self betterment, they interact with other elements of the neoliberal state to create isolation and marginalization among those with low food security, and they fail to value and support the ingenuity and humanity of households experiencing food insecurity. Twenty five years after the rise of neoliberalism and Reagan's cuts to the welfare budget, food access programs in the US continue to be permeated with neoliberal and workfarist subjectivities, which contribute to the creation of a geography of isolation and abandonment that the food insecure are lost in. The book begins by examining the connection between agricultural abundance and food insecurity and argues that overproduction of "industrial food" or "cheap food" (Carolan 2013) counterintuitively leads to food insecurity and hunger as food becomes commoditized and marginalized groups lack the income or power to provision themselves. The book next explores how food access programs often fail to create community among their participants but instead create community among program volunteers and coordinators. Subsequent, using the community economy model developed by J.K. Gibson-Graham (1996; 2006) the book explores the extent to which food access programs discursively construct themselves as alternatives to capitalism and use this embrace of alterity to support their clients. Next, I examine how neoliberal and workfarist regimes are part of the design of food access programs and explore how these organizations interact with other elements of the neoliberal state and describe the geography of isolation created by poverty and food insecurity in the US. The book concludes with an analysis of how food access programs can be restructured in ways that empower clients and meaningfully respond to the epidemic of hunger. In sum, this book is about the relationship between the material overproduction of food, neoliberal social welfare policy, and the lives of people experiencing food insecurity. Given the material overproduction of cheap commodity foods and the rise of workfarism, how can we design a food support system that recognizes the humanity of those in need and empowers those households experiencing food insecurity to advocate for their own self-betterment?

In developing this analysis this book brings together three distinct literatures about hunger and food access in the US: neoliberalism and the rise of workfarism in US social welfare policy (Brenner and Theodore 2002; Peck 2001), the literature on famine and food access that focuses on the political and class disempowerment of the hungry (Watts and Bohle 1993; Sen 1981; Heynen 2009), and the alternative economies literature that focuses on the importance of discursive reframing as a tool of community empowerment (Gibson-Graham 1996; 2006). I argue that food access programs have become a permanent part of the shadow state and are permeated by neoliberal and workfarist approaches to poor people, and work in concert with other aspects of the neoliberal state to disempower communities with low levels of food security. This is especially problematic because, as scholars of famine have argued, food access is a question of political and economic empowerment as opposed to the amount of food production (Watts and Bohle 1993). In fact, overproduction is connected with scarcity as the overabundance of food allows it to be used as a tool of political control. I draw on the alternative economies literature to argue that empowering and pushing back against marginalization and neoliberal subjectivities demands a rethinking of how food access programs are structured. Currently, programs treat clients as deficient individuals in need of support instead of recognizing the diverse ways in which they have agency in their lives. In this sense, programs situated in the shadow state can use their relative autonomy to push back against neoliberalism, however, these efforts are too rare.

This book is based on the experiences of the clients, volunteers, and managers of four different food access programs in Duluth, Minnesota. The four programs are quite different: a traditional food shelf, a food buying club, a service that distributes corporate food donations for a small fee, and an urban agriculture and job training program. It draws on interviews, questionnaires, ethnographic data, and documents from each of the programs. The rising importance of these non-state actors that operate in close cooperation with the state or with populations heavily dependent on state services corresponds with the devolution of poor policy to the local level, and together the case studies comprise ingenious, arduous, and ultimately contradictory spaces within which those in need of food support must necessarily operate. These organizations exist outside of federal welfare policy but serve participants who are in close contact with state social service agencies—and who are often referred to their services by the state. While they each articulate a unique mission statement and way in which they would like to serve people, they operate in a liminal space: constrained by what grants are available, what food is available for distribution, and the logistics of uniting food insecure people with food.

Situating Duluth and the Case Studies

Located at the northern tip of Interstate Highway 35W and about 150 miles north of the Minneapolis Saint Paul Metropolitan Region, Duluth is the fourth largest city in the state of Minnesota. It sits at the western tip of Lake Superior at the end of the Great Lakes St. Lawrence Seaway (GLSLS) and the city's port is the farthest-inland freshwater seaport giving it direct access to international shipping lanes (see Figure 1.1). The city is named for Daniel Greysolon, Sieur du Lhut, a French soldier, trader, and explorer who in the late 1600s is the first European known to have visited the area. Greysolon's explorations of the region came when the area was part of New France, a sprawling colonial project linking Labrador in the East, parts of Central Canada in the West all the way to New Orleans in the South. The economy at this time was based on the beaver pelt trade between the French and native Ojibwa communities. These pelts were traded at forts scattered around the region, and were the cornerstone of a unique hybrid culture that developed as those pelts were harvested in the forests of the Great Lakes area and transported out through the St. Lawrence Seaway for trade in Europe (Taylor 2002; Wolff 1982).

The city's population and economy grew during the early 1900s thanks to its proximity to the Iron Range, the only source of iron in the Great Lakes manufacturing belt. Duluth's population and economy grew because of its strategic location as an essential break-of-bulk point during the US's development as an industrial economy as iron ore was transported to industrial centers across the region. European immigration mainly from Eastern Europe and Scandinavia brought massive population growth to Duluth, and the city adopted the moniker of the Zenith City as civic boosters expected the city to double and triple in size, eventually rivaling other economic centers such as Chicago and New York.

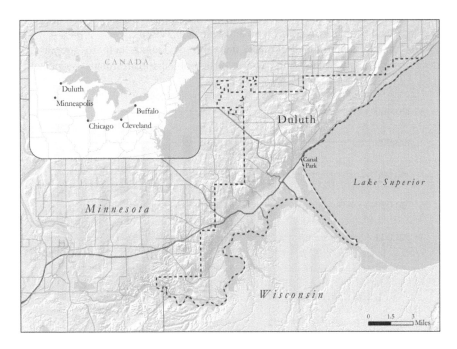

Figure 1.1 City of Duluth.
Source: University of Minnesota Duluth Geospatial Analysis Center.

During this growth period the city was briefly the fastest growing city in the country, and the gross tonnage of its port topped that of major east coast cities.

The trajectory of growth and internal tensions of the city mirrored the rise and fall of other turn-of-the-century industrial centers. Strong labor unions, many with socialist leaning, fought for their share of the fruits of industrial growth, while a smaller group of capitalists who drew their wealth from the lumber, railroad, banking, and shipping industries dominated the city's political class (Hudelson and Ross 2006). Newspapers reflected the multiethnic nature of the population and it was not uncommon to hear Finnish, Czech, and German on the streets of Duluth. In the 1950s, with the decline of iron ore production and the rise of cities outside the Rust Belt, population reached its highest level and from there has declined (see Figure 1.2).

The city is currently dealing with the pangs of deindustrialization and is in the process of repositioning itself as a post-industrial city. The major employers of the area reflect this change: universities, medical centers, and the small airline manufacturer Cirrus are the major employers and unemployment in the city is currently 5.5%, about half a point higher than the statewide rate. The city has a well-developed tourist economy, mostly drawing travelers from the Twin Cities metropolitan areas, where Duluth is both a destination and a midpoint on the way to resorts and cabins on the north and south shores of Lake Superior. Duluth's tourist bubble, Canal Park, is located around the city's iconic lift bridge and packages the role of the port in the city's industrial heritage along with a typical assortment of summer tourist attractions such as carriage rides, ice cream parlors, and local and chain restaurants (Judd and Fainstein 1999).

The unique topography, climate, and size of Duluth help to support its development as a tourist space, but also pose problems for residents—especially

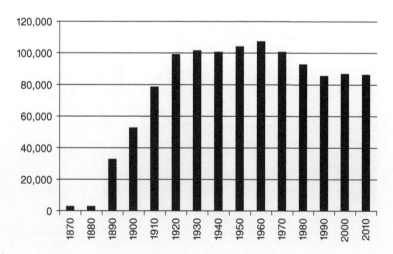

Figure 1.2 Population of Duluth, 1870–2010.
Source: US Census.

those experiencing food insecurity. The older portion of the city—including downtown and the central neighborhoods—sit on a steep slope with elevation changing almost 1,000 feet in the space of a few miles as the lowlands around Lake Superior quickly give way to steep hills. This layout has led to Duluth being nicknamed "the San Francisco of the Midwest," a moniker that is particularly apt because of the neighborhood microclimates created by the maritime climate around Lake Superior and inland climates as you move away from the lake. Duluth's long winters typically bring many sub-zero days with generous snow cover, while summers bring 70F to 80F weather, although it is typically cooler by the lake. The city is slender and averages only three miles in width, but long stretching 23 miles from northeast to southwest along Lake Superior and the St. Louis River Bay. With 86,000 residents, the city covers over 86 square miles creating a challenging climate for mass transit and infrastructure maintenance.

While Minnesota has a history of progressive governance and a comparatively strong social safety net, this support is not all encompassing: only 62% of those eligible for SNAP benefits actually receiving those benefits and 11% of the population experiences food insecurity each year. Duluth in particular has a median income of $35,341 which is 36% lower than the statewide median income and a set of inner-city neighborhoods—East Hillside, Central Hillside, Lincoln Park, Morgan Park and West Duluth—that have areas of concentrated poverty, low-quality housing, and lack of business investment. The needs of these community members are the focus of a range of social service agencies including the large community development intermediary Local Initiative Support Corporation (LISC), as well as social service agencies. While the city is over 90% white, the core neighborhoods are more culturally diverse and house large numbers of African-American and Native American Duluthians. Endemic racism and racial disparities in economic and educational attainment have been long-standing community issues

In terms of civic involvement the city has a very engaged citizenry with strong and politically active unions and a range of food-based organizations. Seventy percent of registered voters participated in the last national election, a full 20 points higher than the national average and school board, county commissioner, and city council electoral forums are often well-attended events. The city's industrial heritage is reflected in the power of its labor unions and Democratic-leaning politics where Iron Range democrats have been a strong voice for blue-collar workers in local and national politics. Within the City of Duluth there is a food co-op with 6,500 members, a strong local gardening community, multiple local farmers markets, and a small community of farmers and food producers advocating for local foods. These farmers and food producers have banded together to create the Western Lake Superior Good Food Network whose goal is to support a healthy, local, just, and accessible local food system, and they support that 20% of food consumed to be locally produced by 2020 (Institute for Sustainable Futures 2013). A recent report by researchers associated with this group analyzed the potential for the region to become self-sufficient in terms of food production, arguing that through land use and diet change the region could feed itself (Stark et al. 2010). The report notes that although in the 14-county area $1.26 billion worth of food is consumed each year, the area produces only $193 million in on-farm production, but a change to local consumption would bring a major economic impact to the region.

Introducing the Case Studies and Exploring Tensions in the Alternative Economy of Food Access

All four of the following case studies acknowledge the problems of hunger and poor food access and have developed a specific system to support those in need. They each acknowledge the inability of their programs to solve the problem of food access, and instead organize their projects around solving the immediate requirements of those in need, and, in some cases, providing other forms of assistance. There is no doubt that although the voluntary sector model of providing food support through charity has been adopted in widespread fashion in the US (Heldke 2009), these civil society projects are insufficient to the scope of the problem and address only a portion of the needs that low-income people have. As Miewald and McCann (2014) write:

> Without a restructuring of the economic conditions that give rise to food insecurity, particularly the absence of predictable, secure, adequate, and long-term state funding for food programs, that charitable food sector, for all the good it does, will remain the fragile frontline of nutrition for the poor.
>
> (552)

But this "fragile frontline" is hardly monolithic: size, culture, theory of change, local political culture, and market forces all play a role in determining the form in which organizations in the civic sector engage in the problem of food insecurity (Wolch 1990; Rubin and Rubin 2007). This flexibility and diversity of form is a hallmark of late capitalism in general (Harvey 1991) and our current food culture in particular (McMichael 2009). Therefore this analysis is sensitive to the variety of discourses and ideologies that form these intentional responses to the entrenched problems of poverty and hunger.

SHARE

Self-help and Resource Exchange (SHARE) was a food-buying club based in the suburbs of Milwaukee. It purchased food directly from wholesalers before turning it over to a volunteer-led network to distribute. The SHARE network reached many parts of Wisconsin, plus Michigan, Illinois, and Minnesota serving 30,000 people a month. In Duluth, about 100 households bought food from SHARE each month. SHARE saved families 30–50% as compared to shopping at a supermarket, and participants chose their food from an order form that includes about 50 items, including staples such as raw meat and chicken, fresh vegetables, processed foods (such as French toast sticks), and fruit juice. Members ordered their food two weeks in advance by phone, online, or by mail, spending an average of $50 per order. The specific products available changed every month depending on what was available, although organizers always made sure that staples, such as fresh vegetables and raw meat, were always available. Food distributions were held once a month in local community centers and churches, and felt both exciting and chaotic as volunteers packed individual orders and members left with their food. In 2013, the organization discontinued service. As a result of the changing nature of the grocery business—especially competition from big-box stores, such as Walmart and Costco, which offered low-cost groceries to a similar demographic—they found themselves unable to find a niche within the broken food system that they were able to serve. Lack of volunteer enthusiasm did not end the project, market forces as well as competition from other food programs—such as Ruby's Food Pantry, which offered even more deeply discounted food than they were able to provide—ended their competitive advantage.

Ruby's Food Pantry

Ruby's Food Pantry (RFP) is a project of Home and Away Ministries, an evangelical Christian ministry based in Pine City, Minnesota, about an hour south of Duluth. The food that they distribute is donations of surplus food from corporations such as Target, Walmart and Gold'n Plump Chicken. Mostly, this surplus food consists of processed foods near their expiration date or limited run foods—like a new flavor of yogurt that is being market tested or a breakfast cereal with a tie-in to a Hollywood movie. In addition, volunteers individually portion food originally packaged for industrial kitchen such as frozen scrambled eggs or pallets of raw potatoes. RFP sells a set selection of these items to customers for a flat fee of $15, which is charged to participants in order to defray transportation costs, although participants receive in excess of $100 worth of food.

Participants usually bring their own boxes to transport their food out to their car or shuttle bus, and the distribution site feels like a festival with volunteers and youth groups from local high schools and other civic organizations handing out the food to recipients. Because so much food is given to each participant, it is difficult for all but the most physically fit individual to carry everything so volunteers often help people out using small trolleys overloaded with food. RFP has become increasingly popular in Duluth, serving around 1,000 people a month at three different distribution sites. It can often take two to three hours for people to receive their food as people check in, receive a number, and then wait for their number to be called. The distribution site is chaotic and exciting as volunteers jovially encourage people to take all the food that they are allotted.

CHUM Food Shelf

The CHUM Food Shelf is located in downtown Duluth and is part of a larger faith-based social service agency that runs a drop-in center, a homeless shelter, transitional housing units, a job counseling program, and a community organizing project. The CHUM Food Shelf plays a vitally important role in sustaining the low-income population of Duluth. In 2013, the food shelf served about 6,000 individuals, roughly 7% of the total population of Duluth. Approximately 35% of their food goes to children. The food shelf purchases about half of its food from Second Harvest Food bank that "rescue[s] food and non-food items from national and regional manufacturers, wholesalers, retailers, restaurateurs and growers." They also receive donations from local churches, food sellers, and purchase government commodity foods. They are in direct competition with RFP for donations of surplus food, although the food bank system is larger and more established and has a much more formalized donation network. While they receive some of the same food as RFP, their distribution method is different. As required by all distributors of government commodity foods, CHUM collects detailed information from food recipients before they are allowed to receive food; these questions address sources of income, income level, and number of individuals in the household. Through this intake process, individuals are referred to other social service agencies for other kinds of support such as subsidized housing and SNAP benefits. CHUM gives away just enough food for the week, and participants are allowed to "shop" for the items they wish to select in a medium-sized room arranged like a small grocery store with grains, vegetables, fruits, and other types of items arranged separately. Recipients are allowed a certain number of points within each food category—depending on family size—and volunteers accompany each recipient as they fill their shopping cart and determine how to spend their points. The atmosphere at the distribution site is clerical as forms are filled out and income eligibilities are determined, and it is somewhat dingy as the actual food shelf is part of a larger complex housing a food donation and distribution site, and other CHUM activities.

Seeds of Success

Seeds of Success (SoS) is a job training program that uses urban agriculture as a way of preparing small numbers of people for the job market. The program is part of Community Action Duluth (CAD), an organization that is part of the national network of community action agencies, which were created in the 1960s as part of the War on Poverty. The mission of these agencies is to foster self-sufficiency in low-income communities. They are funded through federal, state, and local government funding, as well as through grants from foundations. SoS has three goals: job training, community revitalization, and improving access to healthy food. The job training component is accomplished through two cohorts each summer of roughly three members; each member learns the skill of urban agriculture by working for six weeks with a team leader growing, processing, and selling vegetables for sale at a local farmers' market. The community revitalization component is accomplished through farming plots of land in low-income neighborhoods in Duluth. In addition to these scattered-site plots, SoS also has a large facility in a less densely populated area of Duluth. While transporting work crews between work sites is less efficient, maintaining these smaller plots helps address the problem of vacant housing lots. The program works to improve access to healthy foods by allowing program participants to take food home with them, and through their farmers market, which is located in Lincoln Park, the neighborhood where SoS is located, and also a neighborhood often described as a food desert (Pine and Bennett 2013).

All of the organizations exist within the "loose and baggy monster" of the voluntary sector (Kendall and Knapp 1995) and are distinct in terms of what food they distribute, how they fund and perform the labor of food distribution, and how they structure and manage the delivery of food support for participants. These distinctions are important because, as Rose (1999) explores, community organizations offer insight into how community is used as a space of control. He notes the rise of "technologies of community" in which the discourse, structures, and norms at community organizations are used to discipline and control populations. To this end "governing through community involves establishing relations between the moral values of communities and those of individual citizens" (190). As such, participants in community programs become pliable beings upon which technologies of order are directed, and how community groups are organized offers insights into how particular technologies of governance operate. To Rose (1999), the recipient of aid is tied up in the process of social welfare delivery. He writes:

> In the institutions of community, a sector is brought into existence whose vectors and forces can be mobilized, enrolled, deployed in novel programmes and techniques which encourage and harness active practices of self-management and identity construction, of personal ethics and collective allegiances.
>
> (176)

Thus, community organizations are socially constructed and are used to govern from a distance the actions of individuals, and must be understood as strategic places wherein the political subjectivities of volunteers and participants are formed.

In terms of the food distributed, CHUM, RFP, and SHARE each distribute prepackaged industrial food under different distribution models, while SoS grows its own food (see Table 1.1). Thus, while many in the alternative food movement are pushing back against the industrial food system and supporting food that is produced under more ethical and sustainable conditions (Jarosz 2008), it is only SoS that is participating in this project, although they explicitly define themselves as a job-training and community-development program, and not a food systems intervention. RFP charges $15 per portion of food, SHARE sells the food at cost with a small delivery fee, CHUM gives the food away for free, and SoS sells their produce at a farmers' market with steep discounts available for SNAP users. Food quality at RFP depends on what has been donated: there is almost always high-quality frozen chicken, and often bread or bakery products in an acceptable state,

12   The Struggle to Build Community

Table 1.1 Food Distinctions between Ruby's Food Pantry, SHARE, CHUM, and Seeds of Success.

	RFP	SHARE	CHUM	SoS
Price	$15 per "portion:" about $100 worth of food	30%–50% off retail price	Free	Market price, 50% discount for SNAP users
Quality	Middle	High	Middle	Very high
Choice	Low	Middle	High	Medium (limited by the crops produced)
Geographic Availability	Three sites in the Twin Ports	Three sites in the Twin Ports	Three sites in the Twin Ports	One distribution site in a food desert
Temporal Availability	Monthly	Monthly	Monthly	Weekly during growing season

but there have also been poor-quality items available such as frozen pre-cooked scrambled eggs and gallon buckets of hard-boiled eggs. At SHARE, because the food is purchased from a regular distributor, the food is very consistent: high-quality produce, store brand meats, and packaged goods in good condition. At CHUM, food quality varies: there are old non-perishable goods from name brands and store brands mixed with government commodity products, and a small selection of fresh vegetables, which are often not in very good condition and participants spend a lot of time sifting through the goods available to find the exact food they want (Rochester et al. 2011). The SoS produce is of extremely high quality. Choice is very different at the four organizations: RFP clients simply get a portion of food; at SHARE participants select what food they want from an order form; at CHUM participants can select their own food, but within tightly defined categories; and at SoS, participants can receive only produce that has been grown in Duluth during the summer growing season. RFP, SHARE, and CHUM are each at three sites in the Twin Ports area, and SoS distributes at only one location. Similarly, RFP and SHARE have monthly distributions and participants at the CHUM Food Shelf are asked to come only once a month. The SoS farmers' market is open once a week during growing season.

The groups also organize the required labor in different manners (see Table 1.2). RFP and SHARE each have strong leadership teams that manage each distribution site. These teams have periodic contact with paid staff members, but this contact is minimal and they have considerable leeway to develop their sites as they see fit. At RFP, the coordinating committee also organizes the labor of the many volunteers who perform the physical work of the food distribution, while at SHARE the leadership teams manage the distribution site and also perform most of the labor. At CHUM, there is a paid food shelf manager, and she oversees the work of skilled volunteers who have often been at the food shelf for many years. In sharp contrast, SoS has few volunteers; paid crew leaders serve the dual function of job training coach and farm manager. Crewmembers are paid out of job-training grants and are unemployed Duluthians looking for a short-term

Table 1.2 Labor Distinctions between Ruby's Food Pantry, SHARE, CHUM, and Seeds of Success.

	RFP	SHARE	CHUM	SoS
Work Regime	Skilled and unskilled volunteers	Skilled volunteers	Skilled volunteers	Full-time employees and paid "crewmembers"
Gender (volunteers)	Mixed	Mainly Female	Mainly female	Mixed

employment position to make them more desirable when they seek long-term employment. At RFP and SoS, there are a mix of genders, while at SHARE and CHUM the volunteers are primarily women.

The experience for participants at each of the programs is also diverse, and this interaction is crucial because it illustrates the relationship between people experiencing food insecurity and the shadow state (see Table 1.3). Further, since all of the programs draw a distinction between organizer and participant, it is the experience of the participants that defines how the norms and values of the organization are communicated to those in need. RFP and SHARE are each available to all users without eligibility requirement, and the organizations are proud of the universality of their project. At SHARE, the more people use the service the lower the price as they are better able to negotiate bulk-rate discounts, and at RFP upper-income people are encouraged to get a portion to give away to their neighbors in order to make more food available to those in need. In contrast, participants at CHUM must show proof of address and income, and there is an intake process wherein volunteers or the manager talks with them about how they are meeting food needs, and what can be done to discourage food shelf usage in the future such as signing up for SNAP benefits, utilizing WIC, or taking advantage

Table 1.3 Participant Experience Distinctions between Ruby's Food Pantry, SHARE, CHUM, and Seeds of Success.

	RFP	SHARE	CHUM	SoS
Program Eligibility	Universal	Universal	Encouraged to show proof of need	Small pool of job-training clients
Feel	Hectic/ Enjoyable	Hectic/ Enjoyable	Clerical/ Social Service	Social/ Job Training
Time Commitment	Medium	Low	Low	High for crewmembers, low for farmers market shoppers
Level of Need of Participants	Low/Middle	Medium	High	Medium
Stigma Associated with Use	Medium	Medium	High	Medium

of job-training services offered by CHUM. At SoS, only a small group of pre-screened unemployed individuals serve on work crews.

Given the different levels of eligibility, the experience of visiting the sites is distinct: RFP is hectic and crowded as people wait in a large waiting area with hundreds of others for their food, and are then allowed access to a large gymnasium full of excited volunteers handing out food. At SHARE, most participants simply show up quickly to get their food, and the atmosphere is businesslike but cheerful. In contrast, CHUM has a very social service feel as interviews are conducted and eligibility determined, and there is a paternalistic feel as volunteers make sure that only the appropriate amount of food is handed out. At SoS, crewmembers spend eight hours a day engaged in urban agriculture, and the atmosphere is social as crew members and crew leaders talk and laugh, but also pedagogical as participants are expected to perform well in order to receive a good referral and to facilitate their reentry into the job market. RFP works hard to make their distribution non-stigmatizing and encourages volunteers to treat participants with respect. However, given the long waiting time and inconsistent food quality, there is stigma associated with needing to use such a program. Similarly, SHARE works to define itself as empowering and not stigmatizing, but for participants the rigmarole of ordering and receiving food is much less accommodating than a traditional grocery store. CHUM serves those with the very highest levels of food insecurity, and there is a strong stigma associated with food shelf usage. Crewmembers at SoS are viewed by outsiders as employees and therefore not as stigmatized recipients of charity, and the farmers' market looks and feels like a regular, albeit smaller, community farmers' market. In fact, many people speak highly of the quality of produce grown by SoS and look highly upon crewmembers and crew leaders for what they have been able to produce.

Each of the groups has structured the relationship between food insecurity and other anti-poverty programs in different ways (see Table 1.4). RFP operates under the ideology of "pious neoliberalism" (Atia 2012) wherein religious ideology is mixed with an ethos of self-responsibility and community responsibility for social problems. SHARE has its roots in secular community empowerment projects closely linked to community organizers such as Saul Alinsky (1971) and communitarian responses to social problems (Friedmann 1992). CHUM is a traditional charity (Heldke 2009) wherein religiously motivated volunteers raise money and receive grants in order to solve social problems. SoS's organizing theory is based on personal transformation: therefore they work intensively with small groups and involve those individuals in a comprehensive array of social advancement programs (Gray 2010). RFP and SHARE are strictly food access programs and have not branched out into different programs. In contrast, CHUM offers a variety of different stabilization programs across the street from the food shelf at their main building, but users of the food shelf are somewhat distinct from this part of the operation. SoS teaches urban agriculture skills in order to prepare participants for regular employment, and crewmembers are enveloped into the other programs offered by SoS such as money management classes, community discussion, and discounted tax preparation. CHUM and CAD are closely connected

Table 1.4 Ideological Distinctions between Ruby's Food Pantry, SHARE, CHUM, and Seeds of Success.

	RFP	SHARE	CHUM	SoS
Ideology	Pious neoliberalism	Empowerment	Charity	Personal transformation
Role of Religion	Evangelical Christian/multiple congregations	Secular	Multiple mainline congregations	Secular
Range of Service	Only food access	Food access and multiple social services	Only food access	Food access and multiple social services
Relationship to the State	Distant	Close	Distant	Close

to the state and receive a mix of government funding and screen participants for state benefits. In contrast, SHARE and RFP keep their distance from the state, do not receive direct government funding, and prefer not to screen participants for state benefits.

Central to the distinctions between the programs is the way they conceptualize the surplus or commodities that they are distributing and how they define the position of the program participant. At RFP, participants are recipients of God's abundance, encouraged to take everything, and give any extra or unwanted food to their neighbors. It is here that the spectacle of overproduction is most clearly present as participants step back and take a look at all the food and ask "How could there possibly be so much extra food out there?" In contrast, the defining moment of a visit to SHARE is when volunteers tell program participants that they need to double-check their order before they leave because "This is not a charity, you paid for this food and we need to make sure you got what you paid for." Here, recipients are most clearly part of a project that is affirming their status as consumers and empowered members of the market economy, receiving goods at a discounted rate as a result of community organizing. At CHUM, recipients are asked to prove their status as a food-scarce individual, and volunteers watch to make sure that they take only the specific amount of food they are entitled to. Here, participants are recipients of charity and the role of the volunteer is make sure that rules are not broken and too much food is not given out. In sharp contrast to RFP, food is consciously constructed as scarce and therefore must be meted out appropriately to recipients. At SoS, food is produced through the very material process of farming, and the dividing line between industrial food and locally prepared food is made visible.

As Henderson (2004) notes in his analysis of a Minnesota Food Bank, following Marx, commodities are the defining characteristic of capitalist societies, and commodities demand an infrastructure and set of human relationships in order to be distributed and consumed. To this end we cannot delink the different mechanism that delivers food from the ideological and technological processes that produce them. Instead, we must understand them as "interlinked" systems. He argues:

> Food needs human relationships in order to pile up and get around; it gets neither to market nor to the Food Bank of its own accord. It is not immediately apparent to the naked eye what all of these relationships are—'nature' does not determine them—but it does not take much prying to see that the infrastructures for accumulation of food and the mechanisms for its dispersal have become vast and sophisticated. At one pole there are supermarkets, indeed chains of supermarkets, grocery, and convenience stores. At another pole there is a far-flung and socially complex scheme of surrogacy, surrogate labor, surrogate commodities, surrogate economy. There are parallel—no, interlaced—worlds, braided chains of commodification and decommodification. The accumulation of goods is not a pure process.
>
> (489)

In this sense, the distribution of food is one part of these "braided chains of commodification" even as they consciously engage in procedures that reframe and reposition food into something more than simply calories, and people experiencing food insecurity become something more than the hungry.

Methodology

This analysis is based on extensive quantitative and qualitative methodologies at each of the four research sites. Over the past three years I have conducted participant observation at CHUM, SHARE, and RFP. At CHUM, I serve on the organization's board of directors, the food shelf advisory committee and worked a regular shift every-other-week, which put me in regular contact both with other volunteers and the director of the food shelf. At SHARE, I worked every month at one of the three distribution sites in the city until the organizations stopped their services in 2012. At some points I was one of the main contacts at the site and helped to arrange for volunteers and handled the paperwork, and at other times I was part of a team of three to five individuals helping with the distribution. At RFP, I was one of about 50 volunteers that arrived each month to help with the distribution. I was in contact with the coordinating committee, but did not play an administrative role in the organization.

Semi-structured interviews approximately one hour in duration were conducted with 70 individuals involved with the different community-based food programs. At each site, participants, program managers, and volunteers were interviewed. The interview protocol was roughly the same for each group and included questions on their decision to volunteer or use the program, the values they associated with the program, their evaluation of the program, the role food played in their life, their interactions with government and non-profit food programs, and health. All interviewees were offered—and most accepted—a $15 gift card to a local food retailer in exchange for their time. Data was coded using the qualitative research software TAMS Analyzer. All names have been changed to protect privacy, although the names of the organizations are real.

A large survey was also conducted of participants in each of the four programs under analysis. Approximately 300 program participants filled out a 17-page survey, which asked questions regarding their demographic background, level of food security, reliance on government benefits, health, and strength of their community ties. Each person who filled out this survey received a $5 gift card to a local food retailer. The survey was distributed at RFP on one distribution day, at the three SHARE sites in Duluth on one distribution day, and over the course of a month at the CHUM Food Shelf. In addition, some SoS crewmembers also filled out the survey. However, to protect anonymity those surveys are not included in any presentation of data. Instead, they were used to collect some background information.

Finding Community Empowerment amid Abundance and Isolation

This book emerged out of the contention that programs that address food insecurity are not isolated hunger projects but are instead part of the complex shadow state network of voluntary institutions that together form what Jamie Peck and Adam Tickell (2002) refer to as "roll-out neoliberalism." As the state downsizes, what types of organizations emerge, and how should we theorize their actions? The four case studies in this book represent interventions within the community of people experiencing food insecurity that are incredibly varied. Searching within them for a master narrative of food and activism is difficult as each changed and evolved over time in response to internal and external stimuli, and each group has struggled to put their politics and theory of change into action in a meaningful way. Certainly, many are critical of the possibility that civic society can operate in an empowering and transformative way. Janet Poppendieck (1999) argues that an overarching criticism of the food shelf movement is that volunteers spend so much time building community with one another and securing the bare necessities for people experiencing food insecurity that work which challenges the systemic problems of the US food supply never gets done. With the rise of workfarism (Peck 2001) the central goal of the shadow state is to prevent rebellion and instill in the population an acceptance of the status quo. Building on this critique, Nikolas Rose (1999) argues that these programs are not singular but instead plural and have in common a goal of control and discipline. He argues "a whole array of control agencies—police, social workers, psychiatrists, mental health professionals—become, at least in part, connected up with one another in circuits of surveillance and communication designed to minimize the riskiness of the most risky" (260). To accomplish this goal they operate "within the 'savage spaces', in the 'anti-communities' on the margins, or with those abjected by virtue of their lack of competence or capacity for responsible ethical self-management" (259).

Others, however, view the voluntary sector as a more empowering space, one in which neighbors help neighbors overcome obstacles, ethical values are put into place, and the "care economy" is actualized in a way that helps people meet the day-to-day struggle for survival (Cloke et al. 2014). Advancing this argument Gibson-Graham (2006) push back against the totalizing capital-centric narrative of neoliberalism, and instead explore the "practices that are centered upon care of others and the provision of material well-being directly—like the non-market transaction and unpaid labor performed in households around the world (617)" that, along with CSAs, cooperative housing, co-op grocery stores, and the enormous efforts of voluntary organizations and protest movements collectively constitute an "other economy." Read in this light SHARE, CHUM, RFP, and SoS are strategic interventions out of which new subjectivities can be developed and alternative forms of economy and community can develop.

This book critically situates the concept of food insecurity and the organizations provisioning those experiencing food insecurity within ongoing debates on (a) food insecurity in the US; (b) neoliberalism, workfare, and the shadow state; and (c) alternative economies. Food insecurity and diet-related illnesses have reached epidemic proportions in the US, and the US has built an enormous state and parastatal apparatus to address the needs of the hungry that is riven with contradictions. Olivier De Schutter, the UN Special Rapporteur on the right to food, paraphrasing Amartya Sen, writes:

> Unless we take seriously our duties towards the most marginalized and vulnerable, and the essential role of legal entitlements in ensuring that the poor have either the resources required to produce enough food for themselves or a purchasing power sufficient to procure food from the market, our efforts at increasing production will hardly change the situation. For people are hungry not because there is too little food: they are hungry because they are marginalized economically and powerless politically.
>
> (2012, 3)

This book is about just how to engage in this project of addressing the needs of the "most marginalized and vulnerable" in a way that recognizes their need for food, power, and community.

Food Insecurity and Marginalization in the US

Even though the US is one of the largest agricultural producers in the world, food insecurity is present in an astonishing 14% of US households, and 23% of households with children. This points not to a problem with food distribution or inadequate food production, but a fundamental misalignment between the nutritional needs of the population and the way in which food support programs are designed. Those who experience food insecurity are disproportionately marginalized populations: the USDA reports that urban children, children of color, and single-parent households are the most heavily affected by food insecurity (Coleman-Jensen et al. 2013). Heynen et al. (2009) write "parallel capitalist, patriarchal and racist logics that produce hunger and food insecurity in the developing world bear down on vulnerable people [and] cities in advanced capitalist nations" (308). Anti-hunger programs in the US are deeply stigmatizing and run through the model of voluntary programs, charity, and scarcity as opposed to empowerment and community change. Tarasuk and Eakin (2005), for example, note that volunteers at food shelves imagine themselves providing those in need "a helping hand," but this charity-based response to a systemic problem severely trivializes the role food plays in creating life chances for children and families. For volunteers, the sense of community and togetherness they get through volunteering is important to their self-identity (Smith et al. 2010), but this simply reinforces the dichotomy between those with means and those at the margins. Chilton and Rose (2009), critiquing this piecemeal approach to food insecurity, argue that food insecurity should be addressed through a "rights-based" approach that "creates *enabling environments* that support people in nourishing themselves while proving a structure of legal recourse" (1207). This model views not only the short-term need for food as a problem but understands food access as a key component of healthy and nurturing environments, and positions the state as the guarantor (through "legal recourse") of food access.

A key purpose of this work is to adapt the work of food insecurity scholars in the developing world to the US context and situate food access as a political and empowerment question. Watts and Bohle (1993), writing about the dynamics of hunger in the developing world label at-risk communities "spaces of vulnerability" maintaining that hunger is a political problem created by inequality as those with power choose how to allocate food in order to best advance their needs. Those without the class, gender, or political power to advocate for food become victims of hunger regardless of the strength of the local agricultural sector. Shiva and Jalees (2009), in examining the rise of hunger in India, even as the Indian economy has grown, outline the economic policies that support economic growth while simultaneously creating hunger. To whit, there has been an increase in support for large farms, which has increased landlessness, inadequate support for small farmers which has increased poverty, and an increase of imports of cheap food that poor people do not have the money to purchase. In India, specific government policies support wealth creation and unequal distribution of those benefits, and thus hunger. This is also true in the US where neoliberal economic policies have eviscerated urban communities and created rising inequality (Wilson 1996; Reich 2012). While charity-based responses can slow the bleeding, they do not address the underlying causes of poverty and hunger.

Neoliberalism, Workfarism, and the Shadow State

Neoliberalism arose in the 1970s and 1980s as a result of the decline of Keynesianism and the rise of the Reagan/Thatcher administrations in the US and the UK. Neoliberalism is not a one-size-fits-all form of a government but a commitment to class domination and unequal distribution of wealth that is enacted in a broad range of styles. As Harvey (2007) sums up his history of the different ways in which neoliberalism has emerged in South Korea, the UK, the US and Sweden:

> The evidence collected here suggests that uneven development was as much an outcome of diversification, innovation, and competition (sometimes of the monopolistic sort) between national, regional, and in some instances even metropolitan models of governance as it was an imposition by some hegemonic outside power, such as the US.
>
> (115)

To this end the "actually existing neoliberalism" literature reflects the complexity of labeling governing regimes and points out the large architecture of Keynesianism that still exists under neoliberalism and the diversity of state forms under this general governing ideal (Brenner and Theodore 2002). Ever since the rise of neoliberalism there has been a pitched battle to narrowly limit the parameters of state intervention, rollback entitlements, and push work both as an economic necessity and a moral imperative.

The "roll-out" phase of neoliberalism (Peck and Tickell 2003) has been enacted through state policies to discipline and control the population working in tandem with the voluntary sector that pushes the community and the civic society as the appropriate space in which to address societal problems. As Nikolas Rose (1999) writes, "the collective logics of community are here brought into alliance with the individualized ethos of neoliberal politics: choice, personal responsibility, control over one's fate, self promotion, and self government" (249). A key aspect of this is the rise of workfarism which has at its core the creation of subjectivities within which the poor are constructed as deficient, uncontrolled, and in need of proper training in moral comport in order to be ready for work. Thus institutions such as food shelves, the police, social workers, and prisons are strategic sites in which workfarist tendencies and positionalities are constructed. This can be seen in food access organizations through the careful meting of food to clients, classes to teach proper dietary habits, protections against "scamming" by program participants, and the strict dichotomy between volunteer and program participant.

Within the realm of food programs, I argue that this has created a distinct neoliberal bounding of the imagination of food access organizations. Many large programs are structured along the charity model wherein the need for food is depoliticized and groups of volunteers hand out food to those in need. As such, program participants are treated with great care but have little agency and simply become clients of organizations as opposed to empowered actors in their own lives. Although there is leeway within the shadow state for organizations to embrace a wide variety of ways of organizing their programs, neoliberal elements are alarmingly present in each group. Interestingly, regardless of the powerful role that food plays in creating community (Miles 2007), food did not always emerge as a common space where volunteers and participants could enjoy fellowship together. Instead, at CHUM and RFP, volunteers and participants ate different foods and had different relationships with food, and conversations about food between these different groups did not provide a place where community and cross-cultural friendships could be built (for similar arguments concerning obesity see Guthman 2011; Firth 2012; Boero 2010).

The neoliberal bounding of the organizations dovetailed with the larger workfarist regime of the downsizing state to form a "porous continuum of care" and a particular geography of isolation for those experiencing food insecurity. Hungry people must rely on a variety of poorly coordinated voluntary sector and state welfare programs for survival. The struggle to maintain eligibility for these programs, receive food from discounted services, and manage their personal affairs left little time for political advocacy or securing work. Rollinson (1998), commenting on the relationship between homelessness and the operation of the homeless shelter, uses the term "shelterization" to describe the process by which homeless people become routinized into the rhythms, norms, and procedures of the shelter, which are completely distinct from life outside of the shelter. Thus, while workfarism purports to support employment, its very design prevents households from attaining this goal. This new form of "regulating of the poor" tethers their well-being to their ability to manage the diverse requirements of various aid programs and did not give them the opportunity to engage in activities that would alter their economic position (Piven and Cloward 1971).

Alternative Economies

The Gibson-Graham alternative economies project is based on decentering the narrative of capitalism as an all-consuming monolith and instead identifying, supporting, and valorizing the variety of alternative economic patterns that support households and communities (1996; Gibson-Graham et al. 2013). Central to this project is pushing back against neoliberal subjectivities and instead identifying people (and places) abandoned by traditional economic practices as imbued with other values and assets that can be supported in order to help communities prosper (see also Yapa 1996). As such, creating counterhegemonic projects is a communitarian project that replaces individual neoliberal subjectification with new community-based subjectivities. As such, "the community economy is an acknowledged space of social interdependency and self formation" (2006, 166).

Each of the projects in this analysis is engaged in specific practices that decenter capitalism, valorize non-capitalist forms of thinking, and highlight community responses to poverty. CHUM and RFP, in particular, drew on religious theorizations of the relationship between the poor and the economy, and both SHARE and SoS sought to destabilize the neoliberal subjectivities of their participants by creating for them new positionalities as "empowered customers" (at SHARE) and "educated farmers" (at SoS). However, neoliberal regimes were also present within the programs and their alternative economy aspects could best be described as "emerging projects" (Santos 2007) because they merely begin the process of decentering capitalism and its disempowering subjectivities.

As such, this analysis explores how problematic it is to use the voluntary sector as a place wherein alternatives to capitalism can emerge. As Fyfe and Milligan (2003) note, the voluntary sector is a difficult place in which to operate. Larger and more bureaucratic programs tend to mirror state agencies, and organizations are highly dependent on volunteers (who bring their own biases and inequalities to their work, see Musick and Wilson 2008). In an age of widening inequality (Reich 2012), the high level of need found by those utilizing shadow state institutions tends to push agencies towards more clientelization and less empowerment. In this sense, the space of the voluntary sector as a strategic location for interventions into neoliberal subjectivities remains underutilized, but an important place for advancing this project (Rosol 2012).

Chapter Summaries

Chapter 2

In this chapter, I define the literatures used in the book and make three interconnected arguments about food access programs in the US. First, drawing on the work of food security theorists in the context of the developing world, I argue that current food production policies have led to the material overproduction of food. Therefore, the position of people experiencing food insecurity is formed by political and economic marginalization, not by a material lack of food availability. Food access programs must respond to food insecurity grounded in this theorization of overabundance and focus on empowering people experiencing food insecurity and improving their ability to fight for their rights within the existing political and economic system. Second, US food support programs and poor support programs are imbued with neoliberal subjectivities and workfarist ideologies that disempower the poor and do not provide them with the support they need to prosper. The material overabundance of low-cost industrial food is not being used to ensure universal access to calories, instead it is used by the federal government to create neoliberal social policy. This puts the food insecure in a position of being the objects of neoliberal policy as opposed to the beneficiaries of state largesse. Last, I argue that the community economies literature as theorized by Gibson-Graham provides a powerful tool for re-theorizing the overabundance of food and creating new and more empowering subjectivities for food access programs. However, although this literature has been used within the alternative food movement, this body of work has not been applied to the food access literature as fully as possible. Given the stigma associated with food access programs and the overarching neoliberal paradigm present in these programs, this literature provides an important point of engagement for food access programs.

Chapter 3

Here, I turn to the muddy concept of community and explore how relationships developed between participants and volunteers at each of the programs, and the extent to which these connections allowed for participants to feel enough cohesiveness to begin building community economies or to organize collective action for their own self-betterment. Based on the contention that food insecurity is grounded in political and economic marginalization, to what extent are food access program organized in a manner that counters this isolation and supports community? Overall CHUM, SHARE and RFP are places where the volunteers build community with one another, and program users "pass through" on a monthly basis. This pass through helps volunteers make friends with one another and certainly moves food into the stomachs of those in need, but this process highlights how community is formed within community organizations (Lepofsky and Fraser 2003). As McKnight (1995) writes, community is

> the social space used by family, friends, neighbors, neighborhood associations, clubs, civic group, local enterprises, churches, ethnic associations, synagogues, local unions, local government, and local media. In addition to being called the community, this social environment is also described as the informal sector, the unmanaged environment, and the associational sector.
>
> (164)

But the way these spaces and institutions are organized determines how people relate to one another. Because CHUM, RFP, and SHARE are involved in such large logistical enterprises, they are not organized to create any meaningful interaction between participants (Fyfe and Milligan 2003), instead they are organizations that revolve around the critically important task of taking care of those in need of food. Transitioning those people to a more stable position and creating a space for them to come together is simply something outside of their organizational structure and administrative capability.

SoS is unique in that they work intensively with small groups of people (2–3 in each crew), but those individuals really did find both sustenance and fellowship. The physical process of digging in the earth and demystifying the commoditization of industrial food certainly played a role in their ability to build comradeship among participants, but program design seemed more important than the specific tasks they were accomplishing. By working together with participants to accomplish the multilayered project of growing, marketing, and selling produce, crew leaders and crewmembers worked together and created a more interactive space than the other food programs. By employing full-time crew leaders to focus almost one-on-one with participants for a short period of time, it was actually the occasional volunteers who came to SoS that did not participate as much in the sense of community. The small scale on which SoS operates makes it difficult to compare their results across the different case studies, but their experience suggests that accomplishing tasks together helps to bring people together, and limiting the time that volunteers provide for people experiencing food insecurity is key. Instead, making those spaces solely about the needs of program participants is better than making food programs a space for volunteers to come together.

Chapter 4

This chapter builds on the preceding exploration of community in voluntary associations and examines the ways in which this space supports the development of community economies. In a political era dominated by neoliberalism, each of the food programs embrace a mission and sense of being in the world in which they explicitly question and reframe the relationship between food and society. Volunteers from CHUM and RFP see increasing food access as an essential part of their religious identity and are bound up with the project of serving those experiencing food insecurity in a deep and faith-filled way. To them the language of mission and service to the poor is real. Similarly, SHARE has its roots in collective and cooperative action with the poor and draws on the literature of community organizing and empowerment in their program design. For SHARE volunteers designing systems and using innovative strategies to meet people's needs is at their very core. SoS is a project of Community Action Duluth, an agency whose mission is from the War on Poverty, and Community Action is involved in an array of programs whose goal is not to solve the problems of poverty but instead to transform lives.

The transgressive potential of each of the food programs was often less radical to participants than to volunteers and program organizers. Regardless of the strong theoretical underpinnings of the groups, the clientelist form of the organizations worked against a more full realization of the transgressive potential of these spaces. Participants at CHUM, RFP, and SHARE tended to collect their food and move on, and their involvement with the food program was not tied in with a deep engagement with another way of understanding contemporary society and capitalism. One aspect of building a community economy is building community, and this was something that CHUM, RFP, and SHARE built more strongly with their volunteers than their participants. While the actions of these food programs were less than revolutionary, their ideological commitments and identity offer an alternative understanding of how food and community come together and open up spaces for fuller engagement with alterity in the future.

The chapter continues with an examination of the multiple ways in which participants in each of the programs support themselves through active participation in multiple economies. They rely on food from friends and family, hunt, fish, tend vegetable gardens, and view themselves as both users of charitable programs and volunteers who are active members of their community. In this section, I suggest that as these programs grow they can better valorize these alternative subjectivities of self-providers and engaged community members. This is important because, as I analyze in Chapter 5, the concrete manifestations of the workfarist state is so strong that alternative visions of the economy are necessary to provide more liberating subjectivities for program participants. The voluntary sector is a potential space for this intervention, but it must be pursued as part of the overall design and implementation of the program.

Chapter 5

Here, I examine the relationship between neoliberal subjectivities, technologies of governance, and the voluntary sector. At each of the programs, participants are trained to act in particular ways: the are made work-ready, monitored to ensure they take just the amount of food allotted, or made to wait long periods in order to receive support. Understood as part of a system of control and discipline, the format of these programs served to disempower those experiencing food insecurity as they were constructed as deficient and in need of proper guidance in order to survive. The good intentions of the volunteers at CHUM and RFP reinforced these stereotypes as volunteers were placed in a position of policing if too much food was taken, or if an exception to policies could be made for particular participants.

The state sets the macro level food production policies within which food programs operate and also manages the multi-billion dollar poor support programs that many people experiencing food insecurity rely on. Although often understood as separate and distinct sets of policies, overproduction is meted as a form of control, and people existing on the margins carefully coordinate state food benefits, health benefits, and monetary support to create their access to state created surplus. The dominant neoliberal narrative maintains that "the era of big government is over," but for people experiencing food insecurity this could not be less true: they are the recipients of programs like SNAP benefits, WIC, and Medicaid that demand paperwork and proof of eligibility in order to stay enrolled. Maintaining eligibility for these benefits and maneuvering around them was a time-consuming but necessary part of the lives of people experiencing food insecurity. At the level of the individual food program each organization has rules that govern how it operates, and each state program has eligibility requirements that are often complicated and unclear. Taken individually none of the requirements are problematic, but they intersect to create a porous continuum of care within which people experiencing food insecurity must constantly maneuver around requirements without really seeing their place in society change.

Chapter 6

In this concluding chapter, I critically review the data presented in the book and search for ways in which the shadow state can be used to empower and liberate people experiencing food insecurity. I maintain that the voluntary sector is a strategic place within which "spaces of hope" (Harvey 2000) can emerge and where thoughtful interventions against neoliberalism can be launched. However, this can only happen through purposeful program design that pushes back against clientelism and neoliberal subjectivities. Programs in civil society should avoid distinctions between volunteer and participant, and work purposefully to name and recognize the alternative economies that support people experiencing food insecurity. The hungry have intricate relationships with the state and parastatal institutions. Using the limited space of food access organizations to more strongly weave the existing porous social safety net together helps to make these organizations a supportive space, not just another disconnected social welfare institution. Food access is directly related to political empowerment. Therefore, delinking food access with political activity misses an essential place where the relationship between political power and the nourishment needed to live a full and complete life can be advanced.

Looking Ahead

Poppendieck (1999) closes the introduction to *Sweet Charity* with a discussion of the popular Christmas carol Good King Wenceslas, a song about a thirteenth century King of Bohemia whose charity towards the poor of his kingdom and Christian faith made him a national hero of Czechoslovakia. Poppendieck suggests that a sovereign who continues to bask in the glow of his own charitable acts without ever addressing the problems of poverty and inequality in his kingdom is doomed to one day be overthrown by those who would wreak more havoc on the land. She writes:

> This is what might be called the "Wenceslas syndrome," the process by which the joys and demands of personal charity divert us from more fundamental solutions to the problems of deepening poverty and growing inequality, and the corresponding process by which the diversion of our efforts leaves the way wide open to those who want more inequality, not less.
>
> (19)

Echoing this injustice, Mariana Chilton and Donald Rose write: "The existence of widespread food insecurity in a country with the world's largest economy—one that produces a cornucopia of food even to the point of grand-scale exports of surplus commodities—is reprehensible" (2009, 1203).

While Poppendieck focused her research on the singular institution of the food shelf, this book examines food access programs organized in a variety of different models. Kobena Mercer (1992) argues that disambiguation and the politics of difference are key aspects of late capitalism that have helped neoliberal projects embrace the language of difference in the service of marginalizing those out of power. The alternative food movement is involved in the process of creating different economic logics around food (Goodman et al. 2010), but has traditionally been more responsive to the needs and desires of upper-class consumers, and not been responsive to the needs of those experiencing food insecurity. By not assertively articulating a narrative contrary to neoliberalism, programs in the voluntary sector risk being part of the network of workfarist programs embraced by the state. J.K. Gibson-Graham (1996) joke about global capitalism acting as an all-consuming beast, when in fact other economies can and do thrive. They write:

> Marxism has produced a discourse of Capitalism that ostensibly delineates an object of transformative class politics but that operates more powerfully to discourage and marginalize projects of class transformation. In a sense Marxism has contributed to the socialist absence through the very way in which it has theorized the capitalist presence.
>
> (252)

This book is about how we reconcile the important role that institutions providing for those in need are organized in a period where the state seeks to discipline and standardize that behaviors of the population. How do we structure aid in a way that meets immediate needs and empowers individuals to push up against this all-consuming system that opposes transformation?

2 Food Security, the Industrial Food System, and Community in the US

> *"Show me the really wealthy person who has a food security problem and I'll be convinced that food security is a useful unit of analysis. I think that for the most part, for most people, it's a stand-in for a broader economic insecurity."*
>
> Paul, Seeds of Success

Efforts to provide food to those in need have become a ubiquitous and often overlooked part of the fabric of life in the US. To this end Janet Poppendieck has observed that "fighting hunger has become a national pastime" (1999). For example, every year the National Association of Letter Carriers sponsors the event Stamp Out Hunger™, which offers households the opportunity to donate to their local food bank simply by leaving non-perishable foods in a paper bag on their front doorstep that their letter carrier will pick up. Similarly, the Empty Bowl Dinner allows diners to eat food prepared by the finest chefs in the City of Duluth to benefit Second Harvest Northland, the local food bank. Although these programs are everywhere they seem to not be working: Feeding America has a corporate tie-in with *American Idol*, but millions still do not know where their next meal is coming from.

The purely technical question of how to feed the over 300 million people who live in the US has been answered: with the advent of the Green Revolution in the 1940s, food production has increased more than enough to keep up with population growth. Moreover, given the global nature of contemporary agribusinesses expecting a neat connection between domestic food production and caloric needs reflects a naïve state-centric approach to global food production (see for example Abraham 1991). Hunger and food insecurity are not created by insufficient food supply, but are instead created by political and social inequality. Therefore the affluence of the US, counterintuitively, has little to do with the nation's ability to adequately feed all of its citizens. Instead, food insecurity is a problem created by inadequate government response to those living in poverty, how decisions are made about the allocation of surplus production, and the political marginalization of hungry citizens.

Atkins and Bowler (2001), in their analysis of food security in developing countries, argue that poverty, patriarchy, and family size are more closely linked to hunger than food production or acreage of arable land. They maintain that famine should be defined as a "socio-economic phenomena or syndrome" in that famines occur in areas without the economic power, political power, or stability to arrange for food delivery or production, not simply places where the population is unable to grow food (133). Following this argument, Endale et al. (2014), in their study of food insecurity in Ethiopia, found that particular populations within regions experiencing food insecurity are particularly vulnerable to hunger. They write:

> In multivariate logistic analysis, almost all variables like being a female household head, having a large family size, lack of education of the household head, a lack of access to irrigation, lack of income from perennial and off-farm activities and having few or no livestock were significantly and independently associated with food insecurity.
>
> (131)

Similarly, in developing countries, food aid has traditionally benefitted the most vocal or the most politically exigent members of society as opposed to the most hungry (Escobar 1995).

Adapting the views of Atkins and Bowler (2001) and other scholars of hunger in the developing world to the US means that hunger and food insecurity should be understood as symptoms of inequality and political marginalization. Poverty and economic inequality in the US have been growing in recent decades, and the major government and non-governmental programs that provide for these households are not designed to transform people from recipients of aid to economically secure individuals. Instead, they provide just enough food to get by. While rates of inequality were high during the great depression, they fell remarkably during the post-WWII boom years and have steadily risen since the 1970s in a phenomena Paul Krugman (2007) refers to as the "Great Divergence" as high income families—especially those earning income through capital gains and not wage labor—have seen their share of national wealth grow rapidly. According to the Congressional Budget Office the top 10% of households control almost 50% of national wealth (2010). And, while there are more people at risk for hunger, the social safety net has been withdrawn in recent decades through steep cuts in public welfare passed during the Reagan administration and the welfare reform movements of the 1990s (Karger and Stoesz 2009). This has led to higher levels of food insecurity and higher rates of federal benefits usage as more households find themselves unable to provision themselves (Carney 2012).

This high level of inequality and state retrenchment has created the space for an alternative economy to supply food to those in need, one that is riven with contradictions and structural problems. Problematically, while in the past hunger was viewed as one of the problems that poor people suffer from, in the US this linkage has been severed in favor of a parceling off of the symptoms of poverty into distinct conditions which individuals experience (Spencer 2004). These conditions, in turn, create demand for specialized governmental and non-governmental organizations which concentrate on solving these social problems: i.e. "homelessness," "the digital divide," "hunger," and "job readiness," are attacked as distinct conditions, as opposed to symptoms of the same underlying condition of poverty (see Rollinson 1998 for a parallel argument concerning homelessness). The social determinants of health literature, for example, pays special attention to the connection between poverty and health:

> The social conditions in which people live powerfully influence their chances to be healthy. Indeed factors such as poverty, food insecurity, social exclusion and discrimination, poor housing, unhealthy early childhood conditions and low occupation status are important determinants of most of disease, death and health inequalities within countries.
>
> (WHO 2004)

Thus, providing health care while not also addressing the underlying conditions that evolve in tandem with poverty does little to affect overall health outcomes (Farmer 2003). In the context of food insecurity, food is created in such overabundance in the US that simply the overstock from existing sources can cure the immediate sensation of hunger that many Americans face while leaving the structural causes of poverty undisclosed and the highly profitable food production system in place (Tarasuk and Eakin 2005).

Hunger in the US exists at the nexus of the industrial food system, the political and economic disempowerment of people experiencing food insecurity, and the alternative economy of organizations providing sustenance to food insecure households. The first section of this chapter explores food access as a question of political and economic empowerment, and the relationship between federal food aid and federal crop support. Current agricultural policies generously support farmers and the overproduction of commodity crops, such as soybeans and corn, but are paired with poor relief programs that are deeply stigmatizing and offer just enough food to stave off hunger among program participants. Next, I explore the rise of neoliberal economic policy and the concurrent development of workfarist social policy. Pairing the receipt of aid with moral and work-ready behavior further stigmatizes the poor and uses social policy as a way of "regulating the poor" (Piven and Cloward 1971). Further, market-oriented solutions to poor food access and food insecurity, such as farmers markets and buy-local campaigns, use the logic of personal responsibility to address problems created by state policies (Alkon and Mares 2012; Alkon and McCullen 2011; Agyeman and McEntee 2014). Last, I turn to the literature on alternative economies and develop an analysis of food access programs which position them as operating outside of market forces and offering an opportunity for diverse ways of theorizing food access to develop.

Hunger, Famine, and the Political Marginalization of the Food Insecure

The result of the Green Revolution and movements to raise global food production has been a dramatic increase in the global food supply at a rate fast enough to keep pace with global population growth. The UN World Food Programme estimates that even while the world produces 1.5 times as much food as necessary to feed itself, 925 million people experience food insecurity each year (2010; Weis 2007). The ability of global calorie production to meet the needs of the world's population is in sharp contrast to the prevalence of neo-Malthusian claims of population growth outstripping the earth's carrying capacity (Grigg 1985; Ehrlich 1968; Kaplan 1994). These theories rely on the supposition that while population growth is exponential, resource development is arithmetic: therefore population will necessarily grow faster than resources resulting in famines and wars in order to curb population growth (for a fuller critique see Harvey 1974). This theory severely misrepresents the dynamic relationship between economic growth and population growth: population growth is not an independent variable that inevitably rises, but is instead a dependent variable that responds to changes in the organization of society (Abraham 1991). Therefore fertility rates fall as societies empower women, institute social security programs, and create opportunities for their citizens who demand education. More importantly, neo-Malthusian models have an overwhelmingly pessimistic view of global grain production: population growth is occurring concurrent with increased grain production as more land is being brought under cultivation and changes in production techniques have made fields more productive (Abraham 1991; Atkins and Bowler 2001).

A problem with Malthusian logic is that it constructs food access as a simple zero-sum-game of food production, instead of a more complex question of food production and distribution, the allocation of agricultural surplus, and political marginalization. Malthusian logic ignores the political question of how to allocate a non-scarce resource, and instead pretends that there is food scarcity and creates two categories: the hungry "other" (who are at risk of becoming violent and uncontrolled) and the fed "us" (who are safe and civilized). This dichotomy oversimplifies the lived reality of those experiencing food scarcity. People who lack a secure food source in the US may overpay for groceries and shop at local convenience stores or make one monthly trip to Walmart to stock up on cheap calories, thus contributing to long-term health problems, malnutrition, and obesity (Franklin et al. 2012). These strategies solve the immediate problem of hunger but often lead to other long-term health problems, such as diabetes and obesity, complicating a clear delineation between the hungry and the fed.

Michael Watts and Hans Bohle (1993) use the concept of "spaces of vulnerability" to describe a tripartite set of conditions that make people vulnerable to experience famine and hunger: entitlement, empowerment, and class. Drawing on Amartya Sen (1981), entitlement refers to the extent in which individuals can respond to stresses posed by market, socio-economic, and non-market forces to sustainability nourish themselves. Empowerment refers to individuals' experience of powerlessness and the extent to which they can marshal political, class, and gender politics towards their own provisioning. Class refers to the historical models under which food is distributed and surplus is appropriated, and how individuals benefit or suffer from this arrangement. Under this rubric, hunger and famine are created as a result of how groups use political power to advocate for advantageous government policy and how class relationships, provisioning strategies, and level of political influence affect this struggle.

Within the context of spaces of vulnerability, the population of food-insecure individuals in the US can be clearly conceptualized as marginalized members of society whose place within existing gender, class, and economic relations in the US make them unable to provide for themselves within the market. Rates of food insecurity closely mirror poverty statistics in both their geographic distribution and demographic profile: states in the South have the highest rates of food insecurity and poverty, while states in the Midwest and Mid-Atlantic have the lowest rates of poverty and food insecurity. An astounding 27% of Hispanic households, 25% of black households, and 37% of female-headed households experienced food insecurity (Coleman-Jensen et al. 2013). The groups experiencing food insecurity are the same groups of people who suffer from health disparities and diet-related illnesses, indicating a clear link between food access and long-term life chances (Davis and Collins 2014). However, these individuals are spatially dispersed (roughly equal numbers of food insecure individuals live in rural, urban, and semi-urban areas), comprise no simple monolithic ethnic, religious, or dietary group, and are unified chiefly because of their poverty and marginalization from larger society. Hence, using food insecurity as an organizing principle in communities is difficult as this group is so broad. Political empowerment and community cohesion are important components of hunger because food support is connected to protest and community organizing (Piven and Cloward 1971). For example, given the socio-political causes of hunger, the modern IMF/structural adjustment induced "food riot" does not occur in the places suffering from the most severe hunger and famine, but instead take place in cities among organized unemployed workers angry about their inability to purchase food (Walton and Seddon 1994). They write:

> The food riot as a means of popular protest is a common, perhaps even universal, feature of market societies—less a vestige of political industrial evolution than a strategy of empowerment in which poor and dispossessed groups assert their claim to social justice.
>
> (39)

Countries where large urban populations are educated yet unable to find regular work in the tertiary and quaternary sectors of the economy are most apt to riot. As we shall see in the next section, anti-hunger policies are directly related to labor force management and the delivery of welfare benefits as a tool to squelch revolutionary fervor among marginalized and dispossessed groups.

These same spaces of vulnerability exist in the US in spaces that are traditionally referred to somewhat poetically as food deserts in inner-city urban neighborhoods. Food deserts are defined as low-income neighborhoods with poor access to healthy food (Shaw 2006). While many low-income communities have fast- food restaurants and small convenience stores, they often lack full-service grocery stores that stock a range of healthy food options (Black and Kouba 2005). In addition, many residents of these communities do not have adequate transportation and therefore lack the ability to leave their community to shop at grocery stores in adjacent communities. This means that they end up paying more for food while at the same time receiving lower quality food (Blanchard and Lyson 2002; Hendrickson et al. 2006). Lack of access to grocery stores is associated with higher rates of obesity and diabetes (Wrigley et al. 2003; Michimi and Wimberly 2010).

The creation of food deserts has nothing to do with the dynamics of food production in the US and everything to do with the dynamics of urban development. The general pattern of metropolitan form in the US is of fragmentation: the concentration of low-income communities in the inner city and the development of higher-income communities in the suburbs (Dreier et al. 2004). This separation of populations disempowers inner-city institutions, such as schools, as tax dollars are concentrated in high-income suburbs, and dollars are drawn away from inner-city investment towards new infrastructure in the suburbs (Orfield 2002). This fragmentation works hand-in-hand with endemic racism in the US and economic restructuring to create entire areas of metropolitan communities without the education or facility to compete for post-industrial jobs (Wilson 1996). The economic and political situation of inner cities is directly related to provisioning strategies: as retail has followed high-income earners out of the inner city there has been widespread closure of smaller neighborhood grocery stores in inner-city areas (Larsen and Gilliland 2008). This has created a situation wherein car-oriented neighborhoods outside of the city have good access to healthy food, and inner-city communities with limited transit options are cutoff from these new developments.

Neoliberal analyses and solutions to food deserts reflect a focus on the physical absence of a grocery store from a community as opposed to the ability of community residents to purchase food, the conditions under which the food is produced, or the political empowerment of residents of those communities. For example the USDA Food Access Research Atlas (FARA) simply uses buffering of existing grocery stores to identify food deserts, as opposed to more sophisticated analyses of the variable ways in which low-income people access food (Pine and Bennett 2013). As Agyeman and McCentree argue:

> If a supermarket can be located within a USDA-identified food desert census tract, then the residents can presumably count on this as remedying the food desert problem; this logic conveys the message not only that can residents of food deserts *buy* their way out of this problem, but also that the market is responsible for providing a solution.
>
> (2014, 218)

Rather than attack the root causes of poverty and deprivation, locating a supermarket in a food desert is a market based "solution" to a problem that serves to obscure more than resolve the contradictions of US urbanism (Shannon 2014).

Government food policy in the US explicitly supports the overproduction of commodity crops through generous price support programs (Peterson 2009) and the parceling out of meager amounts of food support to the very poor (Ridzi 2009; Poppendieck 2014). While food producers benefit from generous government subsidies that support their continued involvement in agricultural production, low-income consumers are not similarly empowered. Analyzing the relationship between agricultural production and poor relief reframes food access as a political question and positions those experiencing food insecurity as objects of state activity that rarely have the ability to agitate for more favorable treatment.

In the 1930s, the Agricultural Adjustment Administration (AAA) dealt with a glut of pigs produced by American farmers by distributing this surplus meat to the unemployed through what became the Federal Surplus Relief Corporation (FSRC) setting in form a durable connection between the needs of those living in poverty and surplus agricultural production (Poppendieck 1999). The FSRC worked with other Great Society programs distributing surplus foods and household necessities (like coal) to the unemployed (Karger and Stoesz 2009). Although the New Deal era of government anti-poverty programs mainly involved job creation initiatives, such as the Tennessee Valley Authority and the Works Project Administration, these early poor relief programs augmented job creation schemes with commodity distributions efforts. In its basic form the FSRC is still around, although updated and changed through legislative mandates. It is now known as the Food Distribution Division of the Food and Nutrition Service (FNS). As Table 2.1 describes, the FNS distributes commodity food to a wide range of vulnerable populations:

Table 2.1 Main Federal Food Support Programs.

Program	Target Population	Federal Funding (billions)	# of People Served (millions)
SNAP	Low-income families	50	34
WIC	Low-income pregnant women, new mothers, infants, and children	7	9
School Lunch	Low-income students	10	31
Afterschool Nutrition Program	Low-income students	0.3	3
Child and Adult Care Food Program	Daycares and institutionalized populations	2	2
Breakfast	Low-income students	3	11
Commodity Distribution	Low-income Americans	0.5	0.5
TEFAP	Low-income Americans	0.5	~

Source: Food Research Action Center, State of the States Report, 2008.

low-income children and families, tribal governments, the low-income elderly populations, participants in food pantries, and children in subsidized daycare. These distribution programs are run through the United States Department of Agriculture which, rather than the Department of Health and Human Services or the Department of Housing and Urban Development that manage other parts of the federal anti-poverty program, oversees the movement of food from producers to needy consumers.

The War on Poverty in the 1960s brought a second push towards the creation of programs to limit hunger and food insecurity in the US. The Child Nutrition Act of 1966 expanded the range of surplus recipients through the development of Women, Infants, and Children (WIC) program that provides supplemental nutritional support for low-income pregnant women and young children. Along these lines, the federal Food Stamp Act of 1977, now known as SNAP (Supplemental Nutrition Assistance Program), was created; it offers low-income people vouchers that can be used in stores to purchase qualified food items of their choosing. This is the nation's largest and most comprehensive food support program serving the needs of both grocers and consumers.

The 1970s brought a change in the way the federal government provided price supports to farmers. Previous New Deal era policies relied on the ever-ready granary program, land-idling loans, and government grain purchases. The ever-ready granary recognized consumer needs by setting up huge stockpiles of wheat, cotton, and corn. The objective was to carry over the surplus from periods of overproduction to leaner years, thus benefiting the producer during periods of surplus and very low prices and helping the consumer in lean years when supplies would be low and the prices high. This program changed under the Nixon administration in the 1970s to a program that guaranteed a minimum price for corn and provided direct payments to farmers to cover low-prices at market (Pollan 2006). This system encouraged farmers to overproduce as they were fully protected from market pressures, thus creating a permanent surplus of commodity grain production.

The effect of this change was to create a constant oversupply of government subsidized commodity crops such as soybeans and corn. Government, educational institutions, and industry worked together to create outlets for these goods: high-fructose corn syrup quickly became a cheap and ubiquitous sweetener in processed foods, and concentrated animal feeding operations became the standard way for fattening animals for slaughter regardless of the poor ability of cows and pigs to digest corn. As Frances Moore Lappé and Joseph Collins analyzed in their early work on global hunger (1977) "abundance" does not translate into increased food supply as crops are fed to animals in order to produce a more calorie-dense source of protein, and land that could be used to increase crop production is instead used in a less intense manner as cattle-grazing area (see also Shiva and Jalees 2009).

The growth of food production in the US can be understood as both an initiative to protect against famine and food insecurity, and counterintuitively, as one of the forces that have helped to create the conditions for hunger. Considering Watts and Bohle's (1993) theorization of hunger and famine as conditions that exist within a complex nexus of entitlements, class, and powerlessness, surplus grain production has the ability to *exacerbate*, not alleviate, these tensions. Atkins and Bowler (2001) argue that surpluses only survive with state involvement through either protectionist state strategies, or as in the case of the US, price subsidies and direct state involvement in the agricultural process. These processes keep crop prices high in conditions of increased supply that according to traditional economics should result in a lowering of prices. According to data collected by the Environmental Working Group (EWG) crop subsidies between 1995 and 2012 amounted to just under $300 billion (2013). The state has a direct interest in the creation of surplus as a way of protecting farmers from the whims of the market and creating a supply of food that can be distributed as needed by the state. Although we tend to take abundance as a natural (and good) aspect of the food production system, abundance and scarcity are connected to one another at both a theoretical level and a practical level. Food surpluses create hunger as food provisioning is used as another prize to be fought over politically. Under this model, cash is given to farmers under the rubric of "commodity support," thus insuring their political support for politicians willing to protect these subsidies, and create an every-ready supply of surplus commodities for distribution to the needy (Peterson 2009). Some of these commodities are distributed to the US poor through commodity distribution programs, and others go overseas for distribution through USAID as international "charity" or "relief."

Given the countervailing goals of federal food support programs, it is not surprising that food support usage is positively correlated with food scarcity, as those who rely on food stamps receive enough benefits to keep hunger at bay, but not to satiate their need for food or to achieve food security (Nord and Golla 2009; Huffman and Jensen 2003). Rather than promote employment in low-income communities these programs have actually withheld needed support from qualifying individuals through unnecessarily complex application processes, racist relief services staff, and overly intrusive case-management techniques (Piven and Cloward 1971). To this end, social welfare policy is not geared towards supporting marginalized populations, but is instead focused on "managing" the behaviors and subjectivities of the underclass.

Because hunger and food insecurity are symptoms of poverty and marginalization, a key component of food support programs should be the political and economic empowerment of people experiencing food insecurity. Given the socio-political ceases of hunger, how can we design food access programs in ways that address hunger at its root? Food activists, such as Mark Winne (2008) and the Community Food Security Coalition, have embraced the term "Food Security" to describe communities where both hunger and food insecurity are absent and communities are empowered to feed themselves the mix of culturally appropriate healthy foods they desire, acquired in socially acceptable ways. Under the rubric of food security, those reliant on government food subsidy programs are not considered food secure because of their reliance on federal subsidy programs that are unsustainable and socially marginalizing (Allen 1999). While food insecurity and hunger are defined by the lack of food in a community, food security refers to both the presence of adequate healthy food, and the political empowerment to continue to press for an adequate and safe food supply. Expanding on this argument, Gottlieb and Joshi (2010) propose the model of food justice, which they define as "ensuring that the benefits and risks of where, what, and how food is grown and produced, transported and distributed, and accessed and eaten are shared fairly" (6). Importantly, these theorizations of food access push the anti-hunger community to grapple with the long-term and systemic causes of hunger and not merely focus on short-term food access.

Nik Heynen's (2009) work on the Black Panther Party's Free Breakfast program offers examples of the empowering possibilities of food-based activism that attacks the roots of hunger holistically (see also Alkon 2012). Importantly, his work focuses on the Black Panther Party explored how food access was interconnected with other parts of their larger community organizing efforts, not as a means to an end. The Black Panthers in the 1960s realized that inner cities were spaces of vulnerability and began the free breakfast program as a radical response to poverty and deprivation that linked eating to revolution and empowerment (Heynen 2009). Drawing a connection between food and community empowerment, Heynen quotes Eldridge Cleaver (Minister of Information for The Black Panther Party) arguing: "Black children who go to school hungry each morning have been organized into their poverty, and the Panther program liberates them, frees them from that aspect of poverty. This is liberation in practice" (quoted in Heynen 407). The program also challenged the gender norms of the day by having revolutionary men serve food to young children, and put Panther women in the position of both performing traditional "mothering" work as part of the feeding program, but also empowering the children through offering nourishment as part of a revolutionary organization. Finally, the free breakfast program highlighted nutritional deficiency in the bodies of black children and put demands on the state (Governor Ronald Reagan) to institute a free breakfast program itself.

Neoliberalism and Workfarism in US Food Access Policy

Neoliberalism refers to the current strategy of state oversight of the economy that favors deregulation of the economy, privatization of state services, and greater reliance on civil society for the provision of social welfare programs (Harvey 2007; Brenner and Theodore 2002). While Keynesianism favored direct government involvement in the economy, full employment, and demand-side economic policy, neoliberalism focuses on deregulation, "market-based" solutions to the endemic problems of the economy, and supply-side monetary policy. Peck and Tickell (2002) use the term "rollout neoliberalism" to detail the rise of institutions and logics that emerged to govern both the economy and citizens as the state downsizes. Therefore, state retrenchment does not mean that welfare services are no longer provided, instead they are outsourced, privatized, and the political context of their design and implementation has changed. Food shelves are an example of this type of organization as they arose as a specific response to cutbacks in federal food support programs. In concert with deregulation, neoliberalism embraces the rise of civil society organizations to provide services no longer provided by the downsizing state (Wolch 1990; Conradson 2003). Wolch (1990) positions this space as the shadow state, defining it as "a para-state apparatus comprised of voluntary organizations that is outside of democratic control, performs major state functions, yet remain within the 'purview' of state control" (4). Problems with this sector abound: they address problems that have been diagnosed by social service organizations and not necessarily the most pressing problems in particular communities, they are not beholden to the democratic process that oversees state programs, and there is an uneven distribution of resources (Fyfe and Milligan 2003). Organizations become less confrontational and more pragmatic because of their close collusion with the state, and therefore activist or empowering stances are less favored than traditional service delivery programs (Newman and Lake 2006; Elwood 2006; Trudeau 2008).

Frances Fox Piven and Richard Cloward (1971) argue in their groundbreaking work on the politics of poor relief and social support programs that these programs routinely provide less than the needed amounts of aid as a way of promoting work as the only anecdote to long-term poverty as opposed to reliance on the state. In the US, welfare is understood as a form of "poor relief" and comprised of a complex institutional system of government agencies and organizations in the shadow state—such as charities, religious groups—and inter-governmental organizations such as Lutheran Social Services. Piven and Cloward argue that the history of poor relief is "a record of periodically expanding and contracting relief rolls as the system performs its two main functions: maintaining civil order and enforcing work" (1971, xv). Thus, in feudal societies, rioting peasants forced unelected elites to expand social programs, and in contemporary electoral democracies "the votes of an enfranchised populace serve as a barometer of unrest, and the periodic contests for electoral office are intended to exert pressure on political leaders to deal with widespread discontent in the larger society" (40). Therefore, it was the political empowerment of African- Americans in the 1960s after the great migration that forced the development of Great Society programs and an expansion of services for the most marginalized members of society.

Building on the work of Piven and Cloward (1971), Jamie Peck (2001), and Loïc Wacquant (2009b) argue that while periodic expansions and contractions of the welfare state worked as a form of population control during the Fordist era, under neoliberalism the concerted efforts of the enfranchised poor to push for the development of the modern welfare state has been replaced by the emergence of the workfare state. Peck defines this complex institution as "the imposition of a range of compulsory programs and mandatory requirement for welfare recipients with a view to *enforcing work while residualizing welfare*" (10). These requirements are often a combination of activities that are intended to improve the recipient's job prospects (such as training, rehabilitation, and work experience) and those designated as contributing to society (such as unpaid or low-paid work). Given the political nature of protest movements, marginalizing the poor and reducing their impact on the political system is a key policy goal of this ideology.

A clear example of neoliberal and workfarist welfare policy is the 1996 Welfare Reform Act. Under the Clinton administration the entitlement program Aid for Dependent Children (AFDC) was ended and replaced with the Personal Responsibility and Work Opportunity Reconciliation Act (PRWORA) of 1996. While AFDC was a universal program administered at the federal level that provided a baseline level of support for low-income families with children, its replacement—Temporary Assistance for Needy Families (TANF)—imposed a lifetime limit on cash benefits of five years, required recipients to engage in work-related activities as a condition of receiving benefits, and gave states wide latitude in determining how to institute these new policies. Frank Ridzi (2009), in his analysis of Welfare Reform, argues that its passage involved a delicate replacing of the "common sense" notion that the state should take care of its most vulnerable citizens with a new "common sense" ideal that work is the best antidote for the poor. He writes:

> Work-first common sense is as much about destruction as it is about the creation of new and innovative program implementation. In order to move policy and society toward the neoliberal ideas of work-fist the barriers to this model must be undermined and removed. Preparing poor mothers to think about and sell their labor power as a true commodity requires elimination of the family protections that allow them to stay out of the labor market and raise their children. Shifting personal responsibility for economic survival to individuals requires an attack on union protections and entitlements to aid.
>
> (11)

Analyzing the ideological shift in how the poor are viewed as deserving of government support, Loïc Wacquant argues that under neoliberalism there is a constant push towards the limitation of rights for the poor and the use of poor policy to form disciplined neoliberal subjectivities within the poor (2009b). As part of the PRWORA of 1996 Able-Bodied Working Adults Without Dependents (known as ABWAWD to the USDA) are limited to three months of SNAP benefits in each three-year period unless they are taking part in mandatory work or job training activities. Therefore, instead of the state declining in size as a result of neoliberal cutbacks, it is interacting more and more with the lives of the poor, and actually growing through the prison boom (Alexander 2012; Gilmore 2007; Wacquant 2009b), and welfare agencies are pushing the poor to work as opposed to receive benefits passively. The state then becomes a paternalistic power, which "although they might not want to have any part in it, the dispossessed fractions of the working class are the expected 'beneficiaries' of the historic transition from welfare state to punitive state" (Wacquant 2009b, 33).

Shadow state institutions are a key space in which to analyze the neoliberal state: these organizations are autonomous and have flexibility in terms of how they are organized, but instead of empowering individuals to fight for a better position in society, they tend to operate in a disciplining and isolating manner that keep participants dependent on care and unable to advocate for a more secure space in society. As these organizations become larger and more professionalized, they also become less confrontational and more focused on stability.

Fyfe and Milligan (2003) propose a continuum of organizational types with grassroots welfare organizations on one extreme, and corporatist welfare organizations on the other extreme. Grassroots organizations have little formal hierarchy, offer resources for autonomous action, and focus on community empowerment and mobilization. In contrast, corporatist organizations have hierarchies, offer material support, are predominantly state financed, and focus on treatment. As such, many food access organizations view themselves as "service delivery organizations," as opposed to "political advocacy/social movement organizations." This study includes a range of such organizations, from the parastatal corporatist organizations CHUM and SoS, and the more grassroots organizations RFP and SHARE. Brown (1997) notes that AIDS service and activism organizations in Vancouver were filled with the tensions inherent in the shadow state and worked to resist the clientelization of their patrons through discursive, informational, and participatory methods. At the discursive level, they resisted the label "victims of AIDs" in favor of HIV positive. At the informational level, they disrupted their role as informational gatekeeper by ensuring that patrons were aware and knowledgeable of health and policy options available to them. At the advocacy level, they pressed against state agencies to receive better treatment for their patrons. Therefore, larger parastatal organizations can still create alternative subjectivities, not just smaller grassroots organizations.

Consistent with Larner (2000) and Rose (1999), I draw on the Foucauldian concept of governmentality to explore how public policy works—especially through the institutions in the shadow state—as a tool of neoliberal ideologies to control the actions of citizens. Foucault centers his work on governmentality with the idea that governance, thinking broadly, can be defined as "the conduct of conduct" (Burchell, Gordon et al. 1991). Hence, the scope of the state extends far beyond simply enacting laws and managing the economy to a whole set of other cultural and behavioral processes. To Foucault, governmentality is the "art of government" or the "rationality of government;" the study of which seeks to disclose the practices by which the state seeks to imbue in the population the universality of work, self-discipline, and economic independence. Dean (1999) distinguishes the study of government from the study of governmentality by arguing that governmentality seeks to distinguish the particular mentalities, arts, and regimes of government and administration that have emerged since early modern Europe, while the term government is used as a more general term for any calculated direction of human conduct.

Governmentality views the bodies of citizens as the site within which liberalism is transformed from an economic theory to a theory of citizenship (Burchell et al. 1991). In her analysis of micro-credit loan schemes in Peru, Katharine Rankin (2001) maintains that the goal of these programs is to instill in the individual women who obtained credit a new identity as autonomous economic actors responsible for their own welfare. To Rankin, the use of "borrower groups" to instill the importance of loan repayment is consciously designed to remove women from other family and village networks and incorporate them into the larger capitalist economy (see also Bee 2011). The crux of Foucault's analysis of the state is his interest in the idea that, as Ong (1996) points out in her analysis of citizenship, the process of becoming a citizen is one of both self-inscription as well as ascription by the state. On the one hand, the state has an interest in forming citizens who behave in certain economically maximizing ways while, on the other hand, individual citizens have the ability to create their own identity through specific counter-hegemonic projects. Public health is a key part of the citizenship process, as Peterson and Lupton (1996) note, a healthy citizen is one who follows government health advice, is able to engage in economic production, and plays a positive role in their local community. To the state, in sum, healthy bodies are producing bodies. To this end, in a key example of the growth of biopower in the US in the WWII period, supplements were added to milk in order to produce stronger and more able soldiers (Dixon 2009).

In the field of health and food access, neoliberalism and governmentality converge around the ideals of "choice" and individual autonomy in the creation of a healthy lifestyle. Jeanne Firth (2012) notes in her analysis of neoliberal responses to the obesity "epidemic" in the US that

> major food manufacturers (including Kellogg's, General Mills, Kraft Foods, PepsiCo and ConAgra Foods) formed a coalition in 2009 and created the Smart Choices Food Labeling program. The program featured a green check mark label on a wide variety of foods determined to be 'better for you' because they meet Smart Choices nutritional quality.
>
> (39)

Products such as Froot Loops, Cocoa Krispies, and Frosted Flakes got the seal of approval from the panel illustrating the collusion between the façade of choice and the power of industrial food to play a strong role in determining diets. Along these lines, recent efforts to channel federal free and reduced lunch dollars to local farmers as part of the Farm to School (FTS) initiative are touted as a bringing together of state-led anti-poverty programs and local food production. These programs have neoliberal aims as their underlying object: they are consumer-driven processes (in this case, low-income students), and are defended for their ability to raise test scores (and thus produce higher quality citizens) (Allen and Guthman 2006). Neoliberal subjectivities encourage individuals to act as self-maximizing consumers instead of working collectively with other oppressed people to push for better treatment from the state (Holloway and Pimlott-Wilson 2012). As Julian Agyeman and Jesse McEntee (2014) argue, food justice

> has begun to be folded into neoliberalization processes through state involvement and an underlying assumption that food *injustice* can be solved by private market forces, namely the presence of transnational food companies

with increasingly dominant retail arms as well as new types of food provisioning at the geographically localized scale.

(211)

With rates of food insecurity so high, food distribution has become a professionalized aspect of the frayed social safety net of the US. Feeding America is the largest non-governmental entity serving the hungry, and it is clearly a shadow-state institution; it is an almost two-billion-dollar-a-year enterprise with corporate tie-ins with corporations such as Walmart, Zappos, Panera, General Mills and Kraft, and serve 46 million people each year through local food shelves (Feeding America 2010).

While employee run, Feeding America is dependent on volunteer labor. Volunteering is "the unpaid provision of service to others," (Musick and Wilson 2008, 26) and is necessary for the survival of the food shelf system: without the 8.6 million hours of volunteer labor supplied to the system, they would not function (Feeding America 2013). For those experiencing food insecurity, the emergence of "rollout neoliberalism" has meant the partial pullback of the state from direct service provision and the emergence of a plethora of neo-state institutions that provide sustenance to those in need. These organizations vary in form and function, but as Nikolas Rose (2000) points out, most focus on building the capacity for action among clientele that "seek to foster and shape such capacities so that they are enacted in ways that are broadly consistent with particular objectives such as order, civility, health or enterprise" (323). Thus, food education programs and urban agriculture programs are strategic sites where marginalized people are taught proper behavior that is consistent within the neoliberal paradigm (Alkon and Mares 2012; Allen and Guthman 2006). Alkon and McCullen (2011) document the ways in which farmers markets in low-income black neighborhoods both reinforce racist and neoliberal subjectivities, and also provide space for "contestation" of these positionalities to occur. As Eric Holt-Giménez and Yi Wang (2011) write "if the history of U.S. capitalism and social change is a reliable guide, we can be assured that the substantive changes to the corporate food regime will not come simply from within the regime itself but from a combination of intense social pressure and political will" (93). Because of the pervasive nature of the shadow state and the voluntary sector, this is a key space for this political will to arise in.

Alternative Economies of Food Access

Given the enormous discursive and political power of neoliberal economics and workfarist policy, theorizing alternative formulations of the state and the economy is a critical project for scholars and activists seeking to oppose this hegemony. Authors Julie Graham and Kathy Gibson, who wrote collaboratively under the name J.K. Gibson-Graham, have engaged in just this kind of project in their work on alternative economics, which they label the "community economy" and define as a diverse space separate from the mainstream economy in which economic decisions are made in light of ethical discussions and where new languages about the economy and community are constructed (2005; 1996; Gibson-Graham et al. 2013; see also Goodman et al. 2010). To Gibson-Graham, the dominant capitalist economy is not an omnipresent behemoth but instead exists in space along with a myriad of other economic systems that are struggling to come to the fore. They use the metaphor of an iceberg to note that the monetized capitalist economy of private business is merely the visible tip, and it is supported by state enterprises, feudalism, unpaid volunteer labor, and a variety of other uncelebrated economies (2013).

Central to the community economy theory is the assumption that economies are always diverse and in the process of becoming. Therefore, they embrace "weak theory" as opposed to the totalizing theory of capitalism noting that "a weak theory of economy does not presume that relationships between distinct sites of the diverse economy are structured in predictable ways, but observes the ways they are always differently produced according to specific geographies, histories, and ethical practices" (2006, 71). Similar to critiques of "local food," which note the ways in which power imbalances at the local scale are often overlooked (Goodman et al. 2012), Gibson-Graham embrace a critical stance toward community economies, one which understands that an uncritical acceptance of terms, such as local or family, would mask the inequities in these spaces.

Gibson-Graham argue that the creation of new discourses and positionalities are essential for the survival of alternative economies because they decenter the dominant capitalist discourse, and invent a new language for thinking about basic economic concepts such as "surplus" and economic man. They define this as a process of reframing and argue that

> the more we go along with the idea that the economy is an engine that must be fueled by growth, the more we are locked into imagining ourselves as individual cogs—economic actors only if we work to consume. But there are many other ways that we contribute economically.
>
> (2013, 3)

Envisioning the economy as an iceberg explicitly labels household and care work as economic work that should be counted, and provides new subjectivities for individuals to latch onto besides unnamed individual cog or unemployed non-consumer. Part of this reframing is rethinking fundamental aspects of the capitalist economy such as how surplus is distributed and controlled. For example, in cooperative-owned businesses, profit is owned by the collective body of workers that created this surplus, and together they can choose how to allocate this collectively produced benefit. Daya and Authar (2012) note that one important element of community economies is the notion of agency: within the dominant capitalist discourse agency is attributed to "the system" as opposed to individual or collective action.

A central difficulty with launching alternative economy projects is the overwhelming ideological and discursive power of the neoliberal and workfarist state. As Fukuyama (1992) states, neoliberalism labels itself the end of history forcing counter-hegemonic projects to define themselves in neoliberal terms or create a completely new language that will always be under attack. Gibson-Graham, recognizing the dominance of the neoliberal system, use Boaventura de Sousa Santos's (2007) terminology of "emerging projects" to acknowledge that complete freedom from neoliberal discourse is unlikely. Santos argues that "under neoliberalism, the criterion is the market. The total *market* becomes a perfect institution. Its utopian character resides in the promise that its total application cancels out all utopias" (236). Thus, a central project of alternative economies is how to use the space of voluntary associations and the shadow state to propose new subjectivities and visions of utopia that must grow in a context of free-market fundamentalism.

With the state seeking to provide fewer services to underprivileged and marginalized people, groups providing services to these communities have come to play an increasingly important role in poor people's lives and thus play a critical role in articulating the options for an alternative to capitalism to a diverse audience. Levkoe (2011), for example, defines the parameters for a "transformative food politics" as consisting of projects that combine social justice, ecological sustainability, community health, and democracy and outlines projects that—consistent with Boaventura de Sousa Santos's notion of emerging projects—enact some of these transformative politics as opposed to embodying each element. He gives the example of The Stop Community Food Centre (The Stop) in Toronto as a program that serves the material needs of an impoverished community while also creating surplus through sales from their garden, which is allocated towards the needs of the marginalized. The Stop has a variety of different programs including a food bank, drop-in center, farmers market, and urban agriculture program. They use charitable donations as well as social enterprise to fund their food bank, an example of collectivization of resources and surplus being used to support care for the community. As Levkoe notes, The Stop has "worked to develop collective subjectivities within their community, but it has struggled with the continued existence of the food bank as its flagship programme" (699).

One confusing aspect of the alternative economies literature is its connection (both academically and materially) to the various aspects of the alternative food movement that emerged out of consumer and producer dissatisfaction with the industrial food system (Goodman et al. 2010). Out of this dissatisfaction arose new networks and organizations around organic farming, fair-trade networks, support for local agriculture, sustainable food consumption and production, shortening the distance between consumer and producer, and farmers markets (Jarosz 2008). These networks were set up as alternatives to industrial food and engage in a wide variety of distinct interventions. The alternative foods movement has been critiqued for simply adding value to commodities in order to ensure the survival of farmers, while pricing many middle-income people out of the movement (Allen 1999; Guthman 2008). However, many projects in the movement have taken care to avoid elitism and directly involve low-income consumers, such as Will Allen's Growing Power project in Milwaukee (Allen 2013), which—similar to SoS—uses urban agriculture to revitalize urban communities and provide job opportunities for youth while simultaneously addressing problems in the US food system.

Food-based community economies built around individual choice, such as food co-ops and free-trade networks, have been well-studied before (Gross 2009; Little et al. 2010; Dixon 2011; Miller 2003, but for a more critical stance see Slocum 2006; 2007; 2008; Guthman 2008). In contrast to programs that directly address hunger, these are community economies that arise out of abundance in that they require members to pay a certain premium in the form of a membership fee or a higher price for fair-trade certified products. For example, counter institutions, such as farmers markets and fair-trade networks, create new experiences for individuals as economic subjects, and rhetorically reconstruct commodities in the US. Ryan Galt (2013) notes that for CSA farmers in California, the moral economy is a "double-edged sword" as they engage in self-exploitation to support their farms while at the same time feeling personal benefit and enjoyment from taking part in the alternative economy of farm management. In this sense, consumers and producers often have similar motivations for setting up these alternative models and all buy into the overall ideology—for example, eating local or eating with the seasons (Cox et al. 2008), but this is different for systems that are designed according to the client/patron model where unequal relationships rule.

The alternative economy framework has not been well-developed in the analysis of hunger and food-based movements directed at low-income people, but because of the totality of its critique of neoliberal subjectivities it is a critical lens to apply to the civil society organizations that are omnipresent in the life of people experiencing food insecurity. Amanda Wilson (2013) argues that alternative food programs are not nearly alternative enough and argues in favor of a post-structuralist turn in alternative food scholarship which is consistent with Gibson-Graham's embrace of the diversity of existing economic forms and which "questions how particular knowledges and dominant ways of being come to be and seeks to make visible those that have been pushed aside and made hidden by dominant discourses" (726–727). Consistent with Hinrichs' (2000) distinction between *market alternatives* and *alternatives to the market*, Wilson looks to peasant studies and the existence of "autonomous spaces" where those marginalized by economic and political systems are able to provide for themselves based on their own power and communal economies (see Chatterton and Pickerill 2010).

Another example of anti-hunger activism in the context of alternative economies is Nik Heynen's work on Food not Bombs (FNB), which explores how the group drew connections between the militarization of the state and hunger by engaging in public actions that live out its ideology of providing food to all people through offering free food to passersby in elite areas of cities creating unique views of businessmen and the homeless eating side-by-side (2010). To Heynen, this embrace of the politics of survival is a co-optation of food from a tool of social control (handed out to the poor as charity) to a form of mutual aid (Kropotkin 1955[1902]), which has a powerful unsettling force on the existing biopolitics of the US city. As powerful as these symbols are, FNB actions are temporary and theatrical, as opposed to long-term and systemic community building exercises built on long-term sustained efforts to work with those experiencing hunger (Farmer 2003).

Food Access, Community, and Empowerment

The following chapters in this book explore how community-based food projects conduct their work amid this complex and contradictory terrain. The industrial food system relies on generous state support and produces food in overabundance (Patel 2008), yet fighting hunger—often through reallocating this overproduction from one community to another—has become a "national pastime" for those in civil society (Poppendieck 1999). Further, those programs in civil society have been critiqued for not supporting community empowerment (Wolch 1990; Fyfe and Milligan 2003) even as lack of empowerment is a key obstacle to food access (Watts and Bohle 1993). Similarly, the involvement of the neoliberal workfarist state in the lives of disempowered communities revolves around disciplining and policing this population, as opposed to enacting policies that could help support their economic advancement (Peck and Tickell 2002; Peck 2001). While there is a burgeoning alternative food system of programs, such as organic foods and community supported agriculture systems (Jarosz 2008), critics have noted that these programs embody the same neoliberal norms and subjectivities of the conventional food system (Guthman and Allen 2006; Alkon and Mares 2012). Thus, as programs reach out to support those experiencing food insecurity they must carefully position themselves amid these countervailing forces.

A further complicating factor is the ambiguous role of food in the process of community activism. Because of the overabundance of food in the US and the concurrent rise of boutique food movements, such as organic and local diets programs, while there is plenty of food to redistribute to those in need there are also many more voices describing what "good food" consists of and how people should design and manage their diets (Dixon 2011; Pollan 2006; Peterson et al. 2014), and these messages become part of the process of aid dispersal and receipt. Because food is such an important part of how people live and manage their lives, the stigmatizing and moralistic tone of these interactions works against the ability of food access groups to become centers for mobilization (Bell and Valentine 1997; Fothergill 2003). Through innovative uses of community organizing and empowerment, the food justice movement has the ability to create a space to address food access and marginalization at the same time. The tension, however, is the extent to which organizations that serve or provide food access can use food as a unifying and not divisive discursive terrain.

This analysis is sensitive to the difficulties inherent in creating alternative economy projects. Central to this task is creating and articulating unorthodox understandings of key economic concepts and creating spaces where these new discourses can be put into action. Harvey (1991) notes that the flowering of diversity, which arose in the 1960s with the free speech movement, feminist, gay and lesbian rights, and anti-racism work, dovetailed with neoliberalism's support of freedom. To this end, Kobena Mercer (1992) notes "we inhabit a discursive universe with a finite number of symbolic resources which can nevertheless be appropriated and articulate into a potentially infinite number of representations" (427). In this sense projects with a decidedly neoliberal form are pushed under the guise of autonomy and sustainability. The importance of discourse is noted by Alexander (2012) in her analysis of the rise of "The War on Drugs" and concurrent putative measures to combat drug dealers. Such actions arose before the advent of crack and contemporaneous with no discernable uptick in drug-related arrests or drug use. The discursive universe of drug-related criminality bore no resemblance to facts on the ground, indicating the power of representation and social construction over evidence or expressed need. Thus, the task of community-based food programs is to articulate an empowering counter-narrative about the relationship between food and community from the liminal space of the voluntary sector.

The groups in this study occupy a range of positions vis-à-vis the state: while there is close collaboration in some aspects of their programming, there is also flexibility in terms of how they construct and engage in their actions. As such, we can see clear examples of both neoliberal rationales within their programming but also examples of how they use their relative autonomy to enact alternative economies. The relationship between alternative economies and workfarist approaches to poor relief remains an underexplored area of research (Bartlett 2011; Ballamingie and Walker 2013). In the following, I engage with community-based food programs as liminal spaces between neoliberalism and alternative economies, clinging tightly to their desire to improve food access and taking tentative steps towards creating a different future.

3 Creating Community and Empowerment in Community-based Food Programs

> *Most organizations exist,*
> *not for the benefit of the organized,*
> *but for the benefit of the organizers.*
> *When the organizers try to organize the unorganized*
> *they do no organize themselves.*
> *If everybody organized himself,*
> *Everybody would be organized.*
> *There is no better way to be*
> *than to be*
> *what we want the other fellow to be.*[1]

> *By way of analogy, each individual service program is like a tree. But when enough service programs surround people, they come to live in a forest of services. The environment is different from the neighborhood or community. And people who have to live in the service forest will act differently than those people whose lives are principally defined by neighborhood relationships.*[2]

There is an intrinsic connection between food and community: food and restaurant advertising is awash with images of happy people eating together, festivals are often marked by communal meals, and barbecues, brunches, and dinners are key ways in which families and friends come together (Bell and Valentine 1993). Likewise, the slow food movement advocates taking more time to cherish food together, and critiques mainstream fast food for its focus on convenience over quality (Hayes-Conroy and Hayes-Conroy 2010). Churches—which host many food access programs—stress ideas like "breaking bread together" in which food is a symbol of god's love that brings people together (Miles 2007). From a community organizing perspective, scholars of famine and hunger argue that political and social marginalization is closely connected to hunger (Morton et al. 2005; Watts and Bohle 1993). However, as Bell and Valentine note "communities are about exclusion as well as inclusion; and food is one way in which boundaries get drawn, and insiders and outsiders distinguished" (2007, 91). Thus, while food access programs could be places where participants interact and build relationships with one another, using food as a common unifier is not as easy as it sounds. Even though the programs I analyzed were different in design and methodology, my findings are strongly resonant of Janet Poppendieck's (1999) work 15 years ago: for the vast majority of participants, CHUM, RFP, and SHARE food access programs were places to get food but not a space where clients formed friendships or even had a particularly enjoyable time picking out their food.

Finding community within food programs that distribute food is surprisingly difficult; for the participants at these programs one could cynically label them spaces of "anti-community:" at CHUM individuals are interviewed in a process similar to visiting a social worker, and then accompanied by volunteers who ensure they take only the food they are allotted. Similarly, participants of RFP and SHARE get together once a month to pick up their food, but spend little time interacting at the pickup site with one another while they get their food, or learning about the values associated with the program. Many arrive hours early at RFP, but the waiting area feels a little more like the waiting room at a dentist's office than a community festival or social hour.

SoS, which uses agriculture as a form of job training, stands out as a program that connects food access with a strong community-building component. In fact, many participants were so isolated when they originally entered the program that their only close friends were other members of their SoS work crew. The large amount of time that crewmembers spent accomplishing the common task of urban agriculture helped them to build connections with other members of their crew. Likewise, staff spent a lot of time with crewmembers, educating them about the agricultural production process, teaching them how to cook and eat the foods they were growing, and simply hanging out spending unstructured time together driving from field to field or running the organization's food stall at a local farmers market. This time together meant that there were often good feelings and connections between staff and crewmembers. While SoS dealt with a much smaller number of people than the other distributive programs, its impact on the lives of program participants was much stronger and significant.

In this chapter I examine the relationship between community and food access. I ask: *Are RFP, CHUM, SoS, and SHARE structured in ways that create community among their participants? How is clientelization enacted at the programs?* In answering these questions, I make two arguments. First, the structure of SHARE, CHUM, and RFP tended to create community among program volunteers but clientelize program participants. In this way, they tended to act in a manner similar to McKnight's (1995) allusion to the forest in the opening epigraph: they are self-contained systems that serve the needs of volunteers to have a positive social outlet for their volunteer work and meet participants' short-term need for food. At RFP and SHARE, leadership teams formed that transformed individuals into leaders of food access programs, but this sense of ownership and mutuality was not extended to clients. Second, feelings about food were a particular place where differences between program volunteers and program participants emerged. While volunteers at CHUM and RFP shared family meals and enriched their lives through communal eating, clients had a much more instrumental view of food and this difference created tension. This disparity is devastating because program organizers, who came from very different socio-political contexts, tended to have somewhat paternalistic views of the eating habits of program participants (Guthman 2008), and did not always valorize the complex and multifaceted ways that participants used food. The devolution of care work to non-state actors allows for their biases to control program formation, and creates a stasis within the organizations that prevents them from mobilizing for greater social change. SoS—even though they operated at a very small scale—were able to avoid much of the clientelization and created a real sense of community among their clients. Urban agriculture as an activity did not create community so much as the ample space for interaction between program participants and shared sense of vision among program staff. This was a space where empowerment and community were much more visible.

While the concept of community is notoriously murky, most commentators agree that (1) area (physical proximity), (2) common interest (such as identity or specific issues), and (3) social interaction (i.e. meaningful time together) bind groups of people together (Bell and Newby 1978; Laverack 2004; Bell and Valentine 1997). Amitai Etzioni (2004), writing in this tradition argues that communities are

> social entities that have two elements. One, a web of affect-laden relationships among a group of individual relationships that often crisscross and reinforce on another [...] the other, a measure of commitment to a set of shared values, norms, and meanings, and a shared history and identity—in short to a particular culture.
>
> (225)

Because the food-insecure population is spatially dispersed and volunteers tended to live in different neighborhoods from where they volunteered, it is the second two focuses of community that I concentrate on in this analysis: social interaction and a shared sense of values and norms. I argue that the programs were structured in such a way that multi-threaded relationships between volunteers at CHUM—as well as leadership teams at RFP and SHARE—had a chance to build, but for participants at these programs there was not a chance for this sort of integration.

Communities are not natural pre-existing things but are instead socially constructed entities that embody both descriptive and ascriptive elements. As Anderson (1983) writes in his analysis of the national community "all communities larger than primordial villages of face-to-face contact (and perhaps even these) are imagined (6)," and thus while a person may belong to multiple communities a specific process of discussion and nurturance must be used in order to unite people. While Anderson traces this process at the dawn of the modern era among large groups of people, the process of "creating community" for the purpose of community empowerment means identifying and defining a common element (or ailment) and using that shared experience to draw people together (Alinsky 1971). Glen Laverack, writing in the context of health promotion for empowerment notes that "heterogeneous individuals are able to achieve collective action through a process that involves personal action and the development of small groups, organization and networks, in effect the development of community" (45). Hence, a key part of community organizing and community empowerment is the creation of communities based on common interest and need, which can be difficult to identify and unify around in diverse spaces (Young 1990). Therefore, even if program participants do not come in with preexisting sets of multi-threaded connections they can be formed through program design, and programs can be used to build a shared sense of identity to use in the process of mobilization. Thus, creating a sense of community is part of the process of community empowerment, and since this was not seen among the participants in RFP, SHARE or CHUM there was no sort of mobilization among these groups.

A difficulty in creating community within organizations that exist as part of the shadow state is the fact that as the state has withdrawn from the delivery of services to the most vulnerable, these organizations have become professionalized in order to fill this void. CHUM and RFP are serving so many people with such high levels of food insecurity that including any priorities besides the delivery of food was not feasible. Therefore, spaces and times for volunteers and clients to spend time together were eliminated in favor of a streamlining of the distribution process. Further, volunteers perform a form of work: unlike clubs or organizations where everyone is on the same level, there is inherent stratification to relationships at large food access programs (Musick and Wilson 2008; Putman 2000). Volunteers see themselves as delivering services to a population in need and use their identity as "workers" to distance themselves from program participants.

Strengthening Relationships, Building Friendships, and Serving Participants

At each of the distributive food programs (CHUM, RFP, and SHARE) volunteers came together on a regular basis and completed tasks together: sorting food for distribution, setting up the distribution site, helping participants select and carry food, and cleaning up the distribution site at the end of the night. For them, volunteering was a place where community was formed through enjoyable social interactions. Completing these tasks was a source of joy and inspiration in their lives and fulfilled something for them. Kia, an RFP organizer, described the other volunteers at the program in this way:

> I think they all have a good heart and I think they all have pretty much the same type of compassion for people that are really in dire straits or who has those needs, you know, to have people have a healthier food diet or at least give them those options throughout the month.

Volunteering at one of these institutions was to step into a world where service to others and providing a basic human right was so important the volunteers carved time out of their lives in order to make this desire a reality.

For participants, in contrast, each of the distributive programs served their immediate need for food, but tended to do little to attend to their needs for friendship, community, or political awareness that they might need or want. There was no task for them to accomplish to create interaction, there was no shared sense of history or culture between them, and no sense of political activity that was nurtured through program design. Instead, participants consistently described the programs as effectively meeting their needs for food and indicated that their level of food security would be lower were it not for the programs. They also talked about feelings of fellowship and camaraderie at the distribution site but always in less personal ways than the volunteers expressed. As Marisa, a participant at RFP, described her experiences with the volunteers:

> They seat you very comfortably in the church area and they don't question you about the reason you're there. They make it very comfortable for you to be there in the first place, even though you're struggling. That's hard enough. They are very understanding about that. I do like that.

For many, they were grateful that the programs existed, and thankful that the volunteers had dedicated themselves to making the program work—because participants of the distributive programs came from low-income and chaotic lives, and the programs provided a necessary relief to the their difficulties.

As Table 3.1 summarizes, CHUM, RFP, and SHARE served a continuum of people with CHUM clientele the poorest, youngest, with the highest level of unemployment, and the highest levels of food insecurity; SHARE was at the other extreme. CHUM clientele were much more likely to be African-American or Native American, and RFP and SHARE served a majority white group of participants. As the vast majority of the volunteers at all of the sites were white, there was a distinct racial distinction between volunteers and participants at CHUM. Rates of food insecurity were well above the Duluth average for

Table 3.1 Selected Participant Demographic Data from CHUM, Ruby's Food Pantry, and SHARE.

	CHUM	RFP	SHARE
Median Monthly Household Income	$500–$999	$1500–$1,599	$2,000–$2,499
High Food Insecurity	51%	13%	2%
Low Food Insecurity	41%	53%	42%
Food Secure	4%	31%	55%
Unemployed Participants	51%	22%	14%
White	43%	93%	93%
African-American	31%	4%	2%
Native American	28%	5%	3%
Latino	0%	1%	1%
Households with Children > 18	19%	55%	69%
Percentage Female	53%	76%	71%
Median Age	43	45	54

participants in each of the case studies, ranging from close to 100% for CHUM participants to 45% at SHARE. Children were present in the vast majority of SHARE households and two-thirds of RFP families. SHARE also had the highest proportion of female participants.

Caring, Kind, Spiritually Motivated People at CHUM

At CHUM, volunteers have formed friendships and drawn connections with one another through their volunteer work at the food shelf both through interaction and a shared sense of religious values. CHUM operates in a manner similar to the corporatist model defined by Fyfe and Milligan (2003): volunteers are trained and overseen by a volunteer coordinator, much of the financing comes from government sources, and the organization employs case managers and social workers to deliver material support for those in need. Because the food shelf is open five days a week and is always soliciting food, keeping the doors open demands a large amount of effort. Volunteers are needed to stuff envelopes for mass mailings, interview participants coming in for food, sort donations, as well as help participants select food. At-risk youth use the space as a place to do community service, civic groups use it to do community service projects, and, of course, long-time volunteers come on a weekly or monthly basis. A full-time stocker is employed to rotate food and move food from the storage area in the back to the distribution area in the front, and volunteers are always playing a role in making this complex process work.

Interviews with CHUM volunteers tended to reflect the three ideas in the organization's mission statement of "people of faith working together to provide necessities, foster stable lives, and organize for a just and compassionate community" as part of their rationale for volunteering. Consistent with Etzioni's (2004) assertion that a shared culture is part of the definition of community, CHUM volunteers shared a very similar commitment of service to the poor and an action-based theology of service. Linda, for example, a white retired longtime food shelf volunteer, who has played an active role volunteering, has a strong connection to Catholic activism in Duluth and connected volunteering with her desire to set a positive example for her children and make a difference in society. She described her work at CHUM in this way:

> When my kids were younger I used to say to them, "When you make your choice about things you are doing, pretend there are only 100 people in the world, cause you would know all of them and [they would know] all the choices you make, and you would make choices accordingly." So that's what it is. You go on out and you help people that can't help themselves, because they can't. You know if you were there and you needed help, then I would go to help you because you just should, you know, does that make sense?

Linda's clear commitment to improve the lives of the most marginalized members of Duluth's community was typical of CHUM volunteers.

54 *Creating Community and Empowerment*

Many of their volunteers and the majority of the board members come from member congregations and tie their commitment to the poor with their religious background. In addition to providing a way to put their faith into action, CHUM allows volunteers to meet other people and socialize with like-minded folks. Penny, a 62-year-old, white retiree talked about her work volunteering at CHUM as a way for her to define herself now that she is no longer working:

> It's nice to be able to say that I volunteer here, and that takes care of that little conversation obstacle. I do enjoy the group of people that I am volunteering with. You know, we have an opportunity to converse and check in with each other, our lives, and joke and laugh, and tell stories and stuff like that. That's really good, and it does feel good to be putting food into people's mouths, that is part of it. I guess I was raised to understand that you should do for others, that you should do something that involves other people.

Similarly, Cindy, an 80-year-old white part-time teaching aid, and Deborah, a 42-year-old white professional, describe their relationship with the other volunteers at the CHUM food shelf:

> Cindy: They are neat people. I mean we hang together and they are just all neat and caring people. Some of the volunteers need to use the benefits of the shelf, some of the volunteers show up sometimes at the homeless shelter, but they are just all lovely, caring, kind, spiritually motivated people...

> Deborah: Well, I have certainly made friends with the other volunteers there are and the crew that we have on Wednesday night, it's just wonderful. So we have had a couple of get-togethers with the folks there, and I really enjoy the folks that I work with. So that has certainly been a new friendship.

Volunteers for the most part enjoyed interacting and working with people who shared their values. The interaction of serving those in need provided the backbone of their community. While McKnight (1995) argues that voluntary labor unites people, it is important to note that these relationships were meaningful for people, but temporary and task-oriented. At CHUM, people come together to complete the task of food delivery, enjoy the time spent together, and feel fulfilled in their role as volunteer, but do not often get together outside of the food program or engage in the type of multi-threaded connections that Etzioni (2004) or Tönnies (1957 [1912]) use to define community. For example, in his distinction between *Gemeinschaft* and *Gesselschaft* Tönnies notes that in traditional *Gemeinschaft* communities there are multi-threaded connections that tie people together, while in modern *Gesselschaft* communities relationships are singular and based on contracts and deliberate rules of interaction. As the shadow state steps in to perform the functions of government, the professionalization of relationships meant that volunteers interacted more as coworkers than family members; Putnam notes "workplace ties tend to be casual and enjoyable, but not intimate and deeply supportive" (2000, 87). Volunteers came from stable and supportive environments, and tended to already live fulfilling lives of travel and family outside of their volunteer work. Few volunteers talked about meeting up with other volunteers in spaces besides the food shelf. Volunteers came from stable backgrounds and used their volunteer work as a way to augment their already strong social support system.

While volunteers came to CHUM to put their faith into action, participants came to CHUM because they were unable to feed themselves and needed the support of others to make ends meet. With high rates of poverty and high need for food, CHUM clientele were typical of people utilizing social service organizations in that they were in clear material need of support. Their relationships with each other were minimal, and their interactions with volunteers were positive, but not life changing. CHUM does not provide a comfortable waiting place amenable to fellowship or socializing, and the actual process of getting food is an individual interaction between volunteer and participant. Similarly, although many children come to the food shelf, the children's play area is uninviting and filled with cast-off toys.

CHUM, like other food shelves, serves those most in need of food aid; however, there is not a space for this shared sense of need to develop into a sense of community. As one participant described the other food shelf users, "Everybody is well behaved and here for the help they offer. They're very nice people." The high level of need of the participants means that participants are in desperate need of aid and not interested in socializing or meeting other food shelf users, disinterested in challenging CHUM in any way, and their primary goal is to get their food and get home or back to work. While grassroots empowerment organizations would focus on creating relationships between low food security individuals (Fyfe and Milligan 2003), CHUM clients were not interested in making connections outside of their material need for food. Sampson, an unemployed African-American single parent, describes his relationship with other food shelf users in these terms:

> Well, I don't have to know them to have a relationship with them, because, I mean, they are here for the same purpose I'm here for: to survive, to eat, to make sure they have enough food on the table for not only themselves, but for their family as well. They are the same as me, [here to] survive.

This high level of need means that food shelf users often already know each other—for instance through staying together at the CHUM shelter across the street or socializing in other venues. Maureen, a 46-year-old unemployed Native American single parent, talked about food shelf participants as already existing as part of the same community or people in Duluth:

> I know a lot of people, because a lot of people go to the CHUM [Center] and hang out. They come here, and Duluth ain't that big. Like I said, it's so friendly that people just tend to know each other, people that do. Even from

going to the casino with that problem, there's also regulars in there. You just know them, you know them by name, they know you by name, it's weird.

This "sameness" of the social situation of food shelf participants combined with their high level of need to make the food shelf a space of provision and aid, but not socializing and community.

Participants tended to view volunteers positively, genuinely respected them, and were appreciative of the work they were doing at the food shelf. Sampson, for example talked about the volunteers as recognizing needs and looking out for the underserved:

> Well, in my opinion, I do believe that they see the needs of the people, what needs to be done, and just so that they won't suffer for lack of food, and I believe that they look at the greater needs of the people that need programs like this, and I think this should keep going for as long as there are needs.

The religious ethos of CHUM, that housing and food are human rights, meant that although the food shelf was rarely festive, it was imbued with a sense of dignity for those who needed help. An African-American single mom raising her grandson who uses the food shelf said: "They feel the same way I do, everybody needs help." However, because CHUM is set up on a charity model, it presupposes power for the giver, and marginalization for the poor. Lisa Heldke (2009) argues that charity is a form of disempowerment because it bestows upon the giver the power to decide the shape and form of aid "because benevolent gifts are just that—*gifts*—benefactors labor under no obligation to include recipients in decisions about the form gifts should take. It us up to them to decide how to extend their largesse (216)." Participants tended to describe CHUM in instrumentalist terms, indicating that they got the services they needed at CHUM and felt positive about their experience, but CHUM did not play a role in their social or spiritual life. They described the volunteers in positive terms, often ascribing to them the exact same motivations and attributes that they use to describe themselves, but did not talk about them in terms of friendship or camaraderie.

The life worlds of volunteers and participants at CHUM were so different that some volunteers took to referring to participants as living in a "different system." This distinction served to distance volunteers from participants but also recognized the clear systems of marginalization and disempowerment that exist in society (Alexander 2012; Reich 2013; Wacquant 2009a). Volunteers clearly saw themselves as members of the mainstream society who understood their volunteer work as important to both themselves as the wider community. CHUM volunteer Linda expressed the sentiment in this way:

> They are working, but they are not just working in our system and some of those people never could work in our system and because they have been out of it for so long, they are never gonna work in our system. You know?

Nobody is gonna hire them, or if they did they are not gonna make enough, they would still be coming to the food shelf... That's how they are surviving.

The observation that CHUM clients lived in a different world exemplifies the distance between participants and volunteers, the corporatist model of the food shelf, and the aching inequality which characterize both CHUM and contemporary US society.

By focusing their services on serving the most marginal members of society CHUM brings together people from different lifestyles. Penny, a volunteer at CHUM, noted that she did not like to know too much about the participants because this would impinge on the professionalism of the program and the confidentiality of the participants. She noted:

> Well, it isn't really for us to pry or to make people feel uncomfortable. I find that other than saying to them "How are you today?" or "Hope things are going better for you, in the future," something like that, there is some silence there. So if people want to say more, if they have more issues, there is nothing I can do for them really other than refer them to someone else. I try to refer them to across the street [to the CHUM center] if they need other services or if they need to talk about the need or something, but you also really don't have lots of time because there are more people [to serve].

The different lifeworlds between participants and volunteers are rarely pierced and individuals enter and leave the space of the food shelf without leaving their world. This distancing between clients and volunteers is consistent with the rise in neoliberalism wherein populations are governed "from a distance," as individuals are managed through the regulatory regime of the voluntary sector and the interactions that this space produces (Larner 2000; Bondi 2005). This issue is explored in more depth in Chapter 5; here, it is important to note that the structure of the agency and the positionality of volunteers and participants worked against any shared sense of togetherness or community.

The distance between volunteers and clients is also a racial barrier: while people of color represent a mere 7% of the Duluth population, they comprised over half of CHUM's clients, and the volunteers are almost exclusively white. Viewed through this racial lens, CHUM reflects a "white socio-spatial epistemology," (Dwyer and Jones 2000) wherein whiteness is the unexplored and normal identity for volunteers, and their values, privileges, and morals are the unquestioned default against which the racialized Other is judged (Frankenberg 1993; Roediger 1991). To this end, "white cultural practices" (Kobayashi and Peake 2000) imbue the interactions between volunteers and participants: as noted, white volunteers are spiritually grounded, charitable, and comfortable in their position in Duluth. In sharp contrast, many of the participants had experience with racism and racial profiling in the city that worked against their ability to see the city or the white-dominated social safety net providers as welcoming or responsive space. Alan, a 47-year-old African-American food shelf user argues:

> In our society, people tend to look at you by the color of your skin. For that person who is getting that kind of action or heat from people like that, it tends to discourage them to go out and try and be a part of society. They require so much of you to be a part of society. Everybody's not equipped to absorb all that the world has to offer or even be a part of it.

This racialized distinction in how the space of CHUM is experienced works against the creation of community between volunteers and participants. Further, race was something that volunteers spoke cautiously about, if they spoke about it at all. Penny, a 62-year-old white volunteer, carefully tells this story about being (in her opinion) falsely accused of racism during one of her shifts at the food shelf after alerting a participant to a discrepancy in which food had been given to an African-American participant:

> So I said to her, "You know we are supposed to limit this to larger families" and the person actually intimated or inferred that I was being racist because she was black and I just walked [away]... You know I was like really shocked. And anyways, I just walked away and then I continued to help the other packer [...] I was really offended and she came around and she said, "You know I am sorry" and I said "You know I wouldn't be here if I have an issue, I mean. I am not, I wouldn't be here" and she said, "I am just having a bad day." But, I think that's one of the places where I have had to really curb my tongue.

Though there are clear racial disparities in food insecurity rates of food shelf usage, the lack of acknowledgment of the racialized nature of space at CHUM worked against the program reflecting the needs of food shelf users. As Slocum (2007) and Guthman (2008) note, food projects are often formed around white understandings of healthy food and appropriate behavior that are at odds with program participants' needs.

Participants at CHUM frequently talked about their religious background, making religion a common bond between volunteers and many participants, but one that is not vocalized or acted upon by the organization. Participants and volunteers worshiped in different places and although religion formed a strong component of volunteer motivation, the dominant secular thrust of the organization meant that this discourse was not utilized in any way. Usually, participants had a different understanding of faith than that expressed by volunteers. Instead of viewing faith as on obligation to serve others, they drew on faith as a visceral part of their life that served as a reservoir to help pull them through difficult life circumstances. CHUM participant Jennifer, for example, tied her Christian faith to her decision to leave an abusive relationship:

> I didn't really start getting religious until I got into an abusive relationship. Out of all my relationships this was the worst. I come to call on Christ to get me to hopefully see Him every day, and not him. If you get what I mean. I

remember one day ... I was cooking something, and around him I always felt like I had to walk on eggshells. You look at him wrong, he going to argue. You look at somebody that he don't appreciate you looking at he going to argue. I just prayed to God, "God, please let me get out of this mess." Every time I would get out, I would find myself back in. This was the last straw. I cried, I prayed so hard. [...]. I went home. We had just started arguing and I left and I woke up and it was like, whooo. It was lifted off of me. I called, I changed my number, and after I changed my number I haven't talked to him for about a year. I started going to church, I started watching church [on TV] ... I started going to church events and everything. Now, I don't want to say it's an addiction for me to go to church and everything, I just enjoy going.

Similarly, CHUM participant Maureen argued that although she did not attend church, praying provided the strength she needed in order to bring herself through treatment:

I'm not spiritual. I don't go to church or anything like that. I believe in a higher power. Obviously, some beliefs have been instilled in me because I do have them. I have to grab them somewhere. I pray. You find yourself praying from treatment. There is a spiritual part whether people want to admit it, realize it or not when you do go through a program like that.

For CHUM participants who are often at very difficult points in their life, religiosity was a reservoir from which they could draw from in order to find the strength they needed to survive.

In comparison to volunteers who often came from stable faith communities where they served on committees, participants were often searching for a church or searching for a way to find strength from above. Participants described themselves as "church shopping" and looking for a faith community in Duluth, or not at a place in their spiritual journey where church was the place for them. As Rochelle described this feeling: "Going to church, reading the Bible or even watching it on TV, I've not done it. Not that I wouldn't, but I've just haven't gotten that far yet." Given the religious background of many CHUM participants and the religious motivation of the volunteers, this disconnect meant that a common sense of community grounded in religious identity was not able to develop.

The "Sustainable Compassion" of Ruby's Food Pantry

Distribution days at RFP are a gigantic affair involving up to 350 participants, and 50 or more volunteers. A semitrailer full of food pulls up early in the day and is unloaded by volunteers over the course of the next five hours. The food is arranged on two parallel rows of long tables that stretch the entire length of the gymnasium attached to a local church. During the unloading there is an enormous amount of interaction and fellowship between volunteers: they discuss how to manage the chaos, meals they are going to prepare with the food that is being offered, what they have been up to since the last distribution, and local gossip from Duluth. As RFP volunteer Amy described the other people working with her:

> It seemed like everybody was having a fun time and even if it was like dividing potatoes, which is kind of horrible. But everybody is talking and you are talking to people that you would probably not normally talk to, but yet in this conversation you are getting to know people and see different things and experience different things. I think it is a great community, you know, getting people involved with other communities and in their own community.

The actual distribution starts at 3:30, but many participants come hours early to get a number and end up waiting in the sanctuary killing time on their cell phones or napping as they wait for the distribution of food to begin. While the atmosphere is more festive than CHUM, there is not much interaction for participants. Judy described friendships that form while waiting for the distribution to take place:

> INTERVIEWER: Have you formed any personal relationships with the volunteers or staff there? Like friendships?
> JUDY: Oh, no. No friendships, just friends while we're there.
> INTERVIEWER: Oh, okay.
> JUDY: Friends of convenience. Especially the girls at the checkouts and at the desks. We're getting to know them. I have gone enough times to develop a personal relationship.

At RFP, community was formed for volunteers in a very similar sense to CHUM: the religious and church-connected volunteers bonded over the concrete and faith-based act of preparing and distributing food. A counterpoint to the "civic malaise" that Putnam (2000) laments, volunteers found an organized and structured place to build a form of community with one another. Most volunteers spent time together once a month and did not form the multi-threaded connections of community that would define Tönnies' (1957 [1912]) *Gemeinschaft* societies, but instead enjoyed the sociality of distribution day building on their shared sense of commitment to providing for the poor. As RFP volunteer Malcolm described these relationships:

> Yes [friendships are formed], depending on how you define it, I don't know if they do anything outside of Ruby's Pantry. So I don't know if they are friends in that sense. I think people who volunteer have this sense of camaraderie. There is a kind of community that is formed.

In contrast, the leadership team met together much more often and although they came from different churches they formed much stronger "affect laden" multi-threaded bonds that Etzioni (2004) views as indicative of community formation. For participants, the experience was less clientelist than CHUM, but there was little interaction and no space or commitment to identifying similarities among food recipients.

Like the volunteers at CHUM, many RFP coordinators talked about faith as a motivating factor in their volunteering, and their discussions of faith were very grounded in deed over word, and the importance of reaching out and helping others. Volunteers were not theologians involved in acts of martyrdom through volunteering: they saw themselves as led to help others fulfill a concrete demand in their life through their beliefs: As Amy describes:

> You are supposed to do good deeds, you are supposed to help your neighbor and if somebody is hungry you are supposed to feed them, you are supposed to do those things as a Christian, and you are not supposed to keep it to yourself. If you have in abundance you are supposed to try and share your abundance, [...] if everybody was doing that then there would be a little bit of a different world.

There was not a single unifying religious motivation for volunteers, instead volunteers like Amy spoke in personal and passionate ways about how important RFP was to their faith life as a space where they could act out their beliefs.

At RFP, religious motivations strongly influenced how volunteers came to the program, and they often spoke about how improving food access was a "perfect" way for them to put their faith into action. While at CHUM terms like "service" and "care" were common ways of describing their connection to the program, at RFP volunteers spoke in a more general sense about the importance of "action" and helping others. The $15 payment that participants put forward for their food and the monthly nature of the distribution mean that the project is maintainable in the long term. Kia, a member of the RFP coordinating committee, used the term "sustainable compassion" to talk about this monthly project of delivering food to people through RFP:

> I mean it's kind of a sustainable compassion, I think, type of thing... that's the term I use, you know you really feel like "Oh gosh it's a Ruby's week, I'm going to be exhausted!" But after you leave it's really kind of exhilarating, you really feel like you've really helped some people.

Raco (2005) notes the emergence of the meta-narratives of sustainable development and neoliberalism, arguing that as these two ideologies come together they create a hybrid discourse within which the projects of "rolled-out" neoliberalism (Peck and Tickell 2002) embody aspects of sustainable development such as a focus on equity, environmental stewardship, and democratic governance. As Kia uses the term, sustainable is applied to volunteers, and RFP is defined as a project that they as a community of volunteers can "sustainably" provide to those in need. In this same vein, Malcolm, a white minister at a local church that supports RFP, responded to a question about the relationship between faith and RFP by arguing that providing for others makes religion meaningful:

> There are so many stories about Jesus in the New Testament about feeding people and, well, I think there is a metaphor, there is something very, very

moving about providing food for people and I know it is not free food, I know we are charging something for that and even so it's very meaningful for me to be able to provide food in that way and do it in ways I hope are respectfully caring of people. My sense is that people feel cared about, that they do not feel that we are just here looking at them and, you know, kind of give[ing] charity, kind of that negative sense of that. We are sort of above it and right up. My sense is very kind and welcoming the folks that come in.

The "sustainable compassion" at RFP deftly combines a commitment to serving the unfed and the active "caring of people" in a way that combines religious experience with social need, deploying faith as "a metaphysical motivator for individuals concerned about those in need" (Hackworth and Akers 2010).

A key aspect of the type of community that was built at RFP was in the interaction that the setting up and preparing of the distribution site demanded. In contrast to the food shelf where volunteers served the most marginalized members of society and where people were working in a place seen in popular culture in general, and Duluth in particular, as a place where only the very food insecure go, RFP served a more food-secure clientele than CHUM, prided themselves on reducing the stigma of other food support programs, and advertised themselves as a place where "people like you" can go to get discounted food. At RFP, organizers often ordered food from the program and some food-insecure people sat on the coordinating committee. While the belief that "everybody comes here" was widespread, it overstated the similarities between volunteers and program participants. At RFP, volunteers often began their description of the universality of the program, but later talked about the differences between volunteers and participants, sometimes using language very similar to how CHUM volunteers describe the participants: with a clear sense of the otherness of the people in line. As Amy notes:

Oh yes, it's difficult trying to be nice and respectful to everyone all the time, and not knowing where they are coming at or what their day has been like or, in not trying to judge them for what they look [like], or how they act, or to know how their ignorance is... so it's hard to keep our tongue in. Sometimes you get some person who just wants to talk and talk... a lot of times it's like a mentally challenged person or you know you can tell somebody who has some kind of disability, and they want to just sit there and talk and talk and you are like... okay, I got stuff to do.

Similarly, Hettie, an RFP participant described the volunteers at RFP in very positive terms: "They are great, they are laughing, looks like they are having fun, it's not like 'oh we have to be here, we hate it here.'" Participants at RFP spoke in glowing terms about the sense of community among volunteers and the extent to which volunteers made them feel at home and helped them in their interactions at the site. Participants told stories about getting help going out to their car, getting advice on how to cook the food being handed out, and generally making the pickup experience enjoyable. However, even though it was the kind of place that everyone went, there were clearly different experiences for volunteers and for those there for food aid.

Getting Things Done Together: Community at SHARE

SHARE distributions are much less hectic and festive than RFP: with fewer participants, volunteering at SHARE was like a cozy monthly community event as opposed to the spectacle of RFP. SHARE is secular, composed of participants with higher levels of food security, and the types of people it attracts are organized problem solvers who want to be part of a community-led effort making concrete change. The motivation to "solve problems" was clear with program organizers and volunteers who did not bring with themselves a strong religious ethic, but instead a desire to analyze and diagnose problems for the purpose of finding a solution. While CHUM was clearly a bureaucratic non-profit organization, SHARE was closer to a grassroots empowerment organization in that volunteers had greater autonomy in their actions and there was less uniformity across distribution sites (Fyfe 2005). Therefore, while the necessity for sharing the labor of the distribution was similar to RFP and CHUM, the shared sense of values of the volunteers came from a commitment to action that arose out of grassroots belief in community-led social change, not faith. Dana, for example, described how SHARE allows her to bring together the multiple aspects of her personality and professional career to help people, and uses the concept of empowerment (Laverack 2004; Rubin and Rubin 2007) to explain how SHARE accomplishes its task:

> I have a nursing background and I was in public health, community health, and most of my role was teaching people how to lead a good life, whether they have accommodations because of disease or any other impairments they might have had along the way. I have that inclination anyway, but I did my graduate work in social development. I wanted to learn how community organizing helped make a better fit for services to a community. That is my marriage between healthcare in its more rigid, traditional sense and community service, and trying to teach people the skills to make changes in their life that fit them and work towards those goals. Whether that be in a family level, an individual level, or a community level, and teach them those skills about how to find resources—how to make resources happen if they're not there, how to work with other people. This is a natural. It has both things in it. It has the health focus, and it has the community service part.

Further, Dana uses the term "resources" instead of "food" to identify what SHARE brings to people, and highlights the flexibility of the network and its ability to conform to the needs of the community. In her mind, the distribution network is a volunteer-led system that is organized around food but could be utilized to provide other things, if logistical challenges to expansion make sense. Problematically, the organization is still set up to facilitate commodity distribution, albeit through a model that better facilitates community capacity for site managers.

Participants at SHARE tended also to be very no-nonsense, and like SHARE organizers they were orderly and logical people who were willing to take the extra step of ordering food in advance in order to save money. As Bethany, a 65-year-old SHARE participant described:

> Well, it sounded like a good deal. Online you can read what you are going to get, what it is going to cost, and kind of compare the prices you would pay in a regular grocery store, and it is quite reasonable and you get a good variety of things as far as the meat went and the produce.

Because volunteers tended to socialize only at the distribution event itself, relationships between volunteers were similar to those at CHUM and built around task completion, pleasant conversation, and doing good in the world. SHARE develops host sites and trains members of the community to serve as local leaders to facilitate the monthly food distributions. Because these programs were monthly, served less people, and were less corporatized than CHUM (and the larger food banking system), volunteering tended to be a monthly affair that revolved around successfully administering the distribution process. Volunteers came together to package food, unpack boxes, make sure every participant was served, collect receipts etc. This was just the kind of work that was amenable to conversation and fellowship among volunteers as hands were occupied, but often mouths were free to talk, which perhaps helped to create the impression that everyone came to SHARE. Therefore, their volunteers tended to be similar to the people collecting food, but many volunteers were from other neighborhoods and saw SHARE as a place to volunteer in the community. Dana, one of the coordinators at SHARE talked about how people tend to belong to one community, but do not come in contact with people from different socioeconomic or age group in real life, but while volunteering SHARE organizers have a chance to get outside of their community. She noted:

> I always think that at least in our culture, we go to high school and there are all these little cliques of like-minded people, little circles and they don't have very much to do with one another. Not much changes when you're an adult. You have your work group. You have your neighborhood. You have your church group. You have these different circles that don't necessarily overlap. Sometimes they do because people have, they live near people where they work with them or go to church with them or whatever their activities are, but by and large, they have these different worlds they relate to. When people go to a charity distribution, what's fascinating is that people come from all across the board. It is like the little melting pot and people form and have formed for years this fellowship, a true regard and friendship with people who are total strangers to one another.

Emma, a 55-year-old white former small business owner completely agreed with this analysis, describing her favorite and least favorite experience as being a time when SHARE was short of volunteers and the two of us did the distribution together about five people short on one of the busiest days of the year. She said:

> I would say the best and worst are the same experience because there was one time it was a holiday. It was very, very busy and lots of orders and there was only one other volunteer there with me and I believe it would have been you. It was a lot of orders and we were counting. One of us was checking people in, and one of us was filling orders and one of them was just checking [them] out. It was so busy and the two of us got it done and we didn't have any mistakes. I would say they were both the same time.

A key benefit of volunteering with SHARE was the thanks that they got from participants. While SHARE users tended to be more food secure than CHUM participants, many were just above the poverty line and were appreciative of efforts to ease their burdens.

At SHARE, volunteers did not see participants as living in different life worlds, however, volunteers did note that through their volunteer work they came in contact with poorer people who were often in some state of economic or social duress. Although this gulf was less present at SHARE than other food-access organizations, it was often not hard to distinguish between volunteers and participants based on characteristics such as dress, type of car driven, and outward signs of health. Because of SHARE's ethos of empowerment, participants and volunteers were somewhat similar in terms of class background and there tended to be less importance on maintaining a veneer of professionalism between volunteer and participant as was present at CHUM. Instead, people's interactions with one another seemed more honest.

Food, Community, and Food Access

While the overproduction of "cheap food" (Carolan 2013) by the industrial food system constructs foodstuffs as nothing more than ubiquitous commodities to be traded on global markets, to individual volunteers and program participants food is something more than this abstraction. Similar to Bell and Valentine's (1997) observation that food can be used to draw social distinctions, at SoS, CHUM, and RFP all of the volunteers talked about having very strong connections to food: organizers often grew their own food, ate healthy diets, chose to be vegetarian or vegan, and enjoyed cooking meals for friends and family. For them, food was part of their identity and social class (D'Sylva and Beagan 2011; Cox et al. 2008). Dinnertime was often sacrosanct and food was a part of their lives to which they had the time, energy, and desire to instill with their particular food-related desires. SHARE stood out as a site where some volunteers had strong connections to food, but the majority were much more utilitarian and strategic about food and talked about strategies of provisioning like buying food on sale, making trips to discount stores to stock up, and cooking food in large batches in order to minimize prep time. For these SHARE volunteers food was a means to end, not a symbol of their lifestyle.

Participants at all four food programs shared the view of volunteers that food was an important part of their life and in interviews talked about the enjoyment they felt cooking meals for their family and sitting down to dinner with others. Peterson et al. (2014) note the long tradition of upper-class critiques of the food habits of low income, noting that these neoliberal admonishments reflect the idea that "healthy eating" is a moral problem, best illustrated by the poor decisions by low-income people to purchase low-quality and unhealthy food. While some volunteers assumed that participants had a low level of cooking skill or enjoyed eating unhealthy foods, interviews with participants found that they did a lot of cooking and thought very strategically about how to provision for themselves in a beneficial way (see Table 3.2). There is a long literature on how to teach healthy eating and cooking (Flynn et al. 2013; Wood et al. 2008; Hoisington et al. 2002), which constructs those experiencing food insecurity as deficient in terms of their ability to cook and manage their household. Participants, similar to many of the SHARE volunteers, talked about their complex food purchasing and maintaining strategies that spoke to the difficulties of cooking, purchasing, and transporting food on a limited budget with a limited amount of time available to nourish a family. For them food was a part of their life, but it was not a defining component of their life. And, because participants tended to have many obligations on their time, their ability to invest in particular diets or styles of food was limited. Food was part of their life, important, but not highlighted in a clear and particular way.

Because food played such a strong role in the lives of volunteers at RFP, CHUM, and SoS it served as a focal point in their lives, unifying different aspects of their being. Dahlia, a 20-something white crew leader at SoS, describes how in her life food is a hobby, lifestyle, vocation, and obsession rolled into one. She argued:

> Food is huge for me. I raise my own chickens for eggs. Although it's not the most financially sound decision I've ever made, it is an incredible experience the eggs… it's fantastic, in a way that I never could have explained [before] about eggs without having it, that that is the eggiest they can ever be, and they taste like butter. It's very interesting that the manure has increased the production of my own yard and garden dramatically in a way that I, again, didn't ever understand.
>
> I'm a very picky eater and I really don't eat any meat, I don't really eat raw fruit, I eat tons and tons and vegetables, I don't eat dairy. I have low wheat intake, so I'm super picky, and so I'm always very interested in what new cooking, new spices, and new ways of cooking certain foods can override [health] issues […]. And I think that that sort of odd enthusiasm about vegetables and the few things that I do eat […] helps with the work crew too because I get super crazy, ridiculously excited over grains and eating raw carrots out of the dirt.

For Dahlia, working with SoS meant a chance for her to be involved in something she was committed to. Like Dahlia, many volunteers talked about food playing an important role in keeping their families together and saw a connection between

Table 3.2 Cooking by Participants in CHUM, Ruby's Food Pantry, and SHARE.

	CHUM	RFP	SHARE
Daily	24%	30%	37%
Every Few Days	28%	36%	29%
Weekly	8%	17%	11%
Monthly	20%	8%	10%
Less than once a Month	6%	5%	4%
Never	16%	3%	6%

the positive role food played in their life and their volunteer efforts to increase food security. Christine, a 57-year-old white homemaker who volunteers at CHUM, for example, talked about sitting down for dinner with her husband every night and lighting candles—"even if it is the two of us I would like to have some candle there because I think we are really fortunate to have the food we do in this country." Food insecurity as a symptom of poverty played a special role in why volunteers take part in these programs because of the important role that food plays in their own lives. Stacey, a 62-year-old retired RN and CHUM volunteer connected food with her overall commitment to volunteer work:

> I chose food because I think food kind of helps, honestly. I think I would have probably, maybe once the kids were older and stuff, picked up some other volunteer stuff within the CHUM organization, but it just caught hold of me and I just continued doing it, and food is easy in a way because we are lucky enough to live in this community and this country where there is a supply of food to people who need it, and in some way food is easy to volunteer because it is something you can touch and give away. You know that makes sense rather than the advocacy work that needs to be done and at the legislation level and all the other things, it is much hard work... And we all share that [need for food] as human beings. We all need to eat and so we all—fortunately many of us—have never known what it is like to be hungry, but I think we can imagine what it is.

The bringing together of personal attitudes towards food with the opportunity to perform charitable labor highlights the privilege associated with both whiteness and the charitable sector. In light of the important role that community plays in the governing (Rose 1999), the particular class-based views of food held by volunteers become the norm and set the agenda for action irrespective of the beliefs and food attitudes of program participants (Pine and de Souza 2014; Guthman 2008), and there is an inherent power dynamic in choosing food as a symbol around which to perform voluntary labor as opposed to setting the agenda for activism collaboratively (Heldke 2009).

In sharp contrast to the role food played in the lives of CHUM, RFP, and SoS volunteers, many SHARE organizers stood out for their much more utilitarian view of food. Dana, for example, would not define herself as a foodie at all. Dana described eating as "necessary refueling" and nothing more, while Rick, a white SHARE volunteer and community activist, described his favorite meal as pasta, fried hamburger and some Worcestershire sauce. For them in particular, and SHARE overall, the focus of their volunteer work was on solving the particular needs of people, not food in particular. Dana described why not being a foodie was not a contradiction in her work at SHARE:

> No, because to me the food is a footnote. To me it's the community service part. I mean that's really the ... that's why I say if this thing evolves into, you know, a transportation network with people who just really want to help people make their lives better, whether that has something to do with food, so there's that transfer. Great! I mean, just like I was saying before this, there's something in me that needs to do something other than my own little life here.

By distancing themselves from the food movement (Pollan 2006) and embracing a more utilitarian relationship with food, SHARE avoided the moralistic and paternalistic messages around food that formed some of the food talk at CHUM and RFP, and allowed food to be more of a common ground among volunteers and participants.

In this vein, participants at all of the programs spoke about food in instrumental terms and described the often sophisticated methods that they used to purchase and ration the food that they ate. They viewed food as something that was important to them and their lifestyle, but for them food was a part of their life that had to be managed in the same way with which they dealt with other aspects of their life. For example, participants talked about couponing, shopping at lower-priced grocery stores, stocking up on items that were on sale, getting higher-cost food items—such as meat from the food shelf—and supplementing this from purchased foods, and cooking in bulk to make their food dollar stretch. Organizers at CHUM and SoS went to great length to produce and consume food that was consistent with their political and social ideals. This embrace of food as an organizing principle coheres well with Alkon and McCullen's (2011) analysis of the racial coding of farmers' markets as white spaces of privilege. As they write, "farmers markets such as our cases become inclusive, empowering spaces for a form of food politics that reflects liberal, affluent, white identities and positionalities" (939) that construct a white community and actively works against the inclusion of others outside of the clique of privileged consumers. In contrast participants tended to view food as a necessity which subject to the same constraints as other aspects of their life. Therefore, conversations with food program participants revolved around their strategic use of the food available to them, and how they were able to manage food in a way that also allowed them to reach other goals in their life.

Sampson, for example, a 42-year-old, African-American food shelf participant, talked about how he managed the food in his possession by making sure to "ration out" everything in the order it was purchased in order to avoid spoilage or food waste. He described the process by noting:

> Okay, I go by rotation. Rotation is simply that whatever I get, I supplement it with what I already have. I eat what's already been opened. Let's say I went shopping on the 10th of May, I would eat what I purchased on the 10th of May, and whatever I would get the day later I would put that aside and so on to rotate my food so that we won't go through food too fast like I see my stepdaughter's kids doing.

Sampson argued that it was a "sin before god" to waste food and carefully managed what was available to him to avoid this problem. Jennifer, a 48-year-old, unemployed African-American CHUM participant talked about how she would strategically ration out frozen food in order to make sure there were no leftovers (which she did not like to eat) but still have fresh food for her to eat:

> If I get a small thing of hamburger, I might use a sandwich bag or something to divide that up and take it, thaw it out, and then use a little potato that I got. I just make a portion just for one person. I'm sort of good at stretching until I can actually go to the grocery store or whatever.

By carefully managing the food available to them, food-program participants were able to stretch their food dollars and ration out the food available to them. Sampson and Jennifer's careful and strategic protection of their food supply pushes back against portrayals of the food insecure as unintelligent utilizers of food, or as individuals who engage in what Anater et al. (2011) refer to as "unsafe food practices." This finding confirms Whiting and Ward's (2010) careful analysis of the multiple provisioning strategies of the extremely food-insecure residents of the Northern Cheyenne Indian Reservation. I examine these processes more in Chapter 5, but my interest here is arguing that program participants had a different—but no less important—relationship with food than volunteers. However, their views were not in line with the central tenets of the good food movement such as local food and the sacredness of food (Pollan 2006) and instead reflected an intelligent yet pragmatic organization of their food resources.

Participants also used their cooking skills as a way to stretch their food dollars in ways that pushed back against dominant cultural constructs of low-income people as wasteful or irresponsible (see for example Ridzi 2009). By cooking food in bulk, buying cheaper cuts of meat that could be slow-cooked, or cooking dried beans, participants were able to stretch their food budgets. As Katrin described her process:

> Instead of making small portions, I buy the freezer bags when I go grocery shopping. I make large portions like beans; when I boil my beans I make enough that I can just freeze them. Or if I make spaghetti, I make enough that I can portion it out.

A commonality among all of these strategies was that they took time and effort from participants. Participants approached community food program as strategic locations, which they approached intentionally as places that could sustain themselves and their families in a safe and healthy way. In this sense, healthy diet interventions that teach food shelf users how to cook and purchase low-fat ingredients (for example Flynn et al. 2013) should only be implemented in ways that respect the existing knowledge held by food shelf users.

A key place where the different understandings of food held by volunteers and participants was the advice that volunteers often gave to participants about what food they should select. CHUM is set up on the "participant choice" model wherein participants have the ability to walk through an ersatz grocery store where volunteers will make sure that they select the appropriate amount of each type of food, and offer advice as to what food they should select. As volunteers closely monitor each item that is selected, they are recording a list of items that define who the shopper is. As Cramer et al. (2011) write in their book on communication and food: "we often use food to communicate with others and as a means of demonstrating personal identity, group affiliation and disassociation, and other social categories, such as socioeconomic class" (xi). RFP has so much food available that inevitably some participants do not recognize certain foods, and this becomes a place where volunteers can teach participants how to cook or serve unfamiliar food, or volunteers can steer participants towards healthier choices. Some volunteers encourage participants to select healthy vegetables, while others choose to allow participants to choose the food that best suits their interests.

These conversations about food choice were often brief and informal, but had the potential to be permeated with the very different relationship with food that participants and volunteers had. Food selection is a deeply personal and telling process: it illustrates your lifestyle, personal tastes, level of cosmopolitanism, level of cooking skill, and also the types of chronic health problems that you have. Volunteers therefore witness which participants take fresh vegetables and which do not, which take ice cream and which choose not to in a public and inevitably judgmental atmosphere. Some volunteers viewed participants as having a low level of cooking skill and saw them as unfamiliar with different types of fruits and vegetables that were available at RFP and CHUM. In contrast to many of the volunteers who are avid gardeners and cooks, they did not know about or understand the set of rationing skills that participants possessed, and instead looked down on their perceived lack of cooking skills. Stacey, a CHUM volunteer, neatly combined the arguments of Peterson et al. (2014) and Vileisis (2009), and the social distance between volunteers and clients expressed by CHUM volunteer Linda:

> There is a whole generation of people that have grown up that have never experienced [a home-cooked meal]. They don't know what cooking at home is except for Thanksgiving, maybe. It doesn't happen. This isn't money, plenty of wealthy people do not sit down at a meal together so it isn't money, it's that people aren't doing that. A meal, preparing a meal, isn't a big thing anymore. Or it's packaged or it came from the restaurant or got picked up on the way home, well that isn't, there is no prepping in that.

Volunteers tended to view participants as unfamiliar about foods and in need of instruction about how food should be cooked and handled.

Volunteers and participants could best be described as coming from very different food worlds, which means they often have difficulty communicating about food issues. Caring about food is something that comes very naturally to the volunteers: they spend time researching and planning menus, have backyard gardens, and devote their limited leisure time to nurturing this interest. Participants, in contrast, care about food supply and access, and also devote considerable time to the very different act of procuring and allocating their scarce supply of food. Because of their middle-class status, volunteers do not view or understand the trajectory and web of decisions they have made about food in anything similar to the ways in which the participants have done, and through their volunteer work they interact with people whose relationship with food is very different from their own. As Fielding-Miller et al. (2014) found in their analysis of the concept of healthy food in the context of food security among HIV-positive sex workers in Swaziland:

> Sex workers who are living with HIV spoke positively about the clinical advice they had received to "eat healthy foods," but the definitions of "healthy foods" ranged from having any food available at all, to refraining from alcohol abuse or fried foods, to integrating more fruits and vegetables or more traditional Swazi foods into their diets.
>
> (81)

The largely middle-class volunteers guided people towards particular food that reflect their understanding of the term 'healthy,' but did not always valorize the complex and innovative systems that food-insecure individuals utilize in order to stretch their limited food dollars.

Digging for Community: Urban Agriculture as a Place for Community Formation

Arriving at SoS on a cool and overcast summer morning is like stumbling into a covert school of urban agriculture where course timings and class offerings are a closely held secret: the front desk has no idea where the SoS crew is located and I'm sent down a hallway in search of the loading dock. CAD's office is in an old elementary school, and the SoS crewmembers are often in little corners of the building working on projects. One group is in an old classroom watering seedlings, while another is in an underutilized kitchen cleaning beets for sale at the farmers market. Paul, a crew leader, is loading supplies into an old van, and the crewmembers are discussing how many lots they will make it to today, if it will be possible to use a weed whacker on such a damp day, and where a chronically late crewmember is. SoS has a small number of crewmembers who work intensively with one another for a short period of time and they quickly form a close-knit community with one another. This experience of urban agriculture bringing people together has been found in other analyses of these types of programs (Carney et al. 2012), and could be seen as similar to Community Supported Agriculture programs where members work on the farm together and build community through this time together (Cox et al. 2008). SoS is clearly located in the shadow state and is a corporatist social welfare organization (Fyfe and Milligan 2003). However, they use their comparatively large budget to create an atmosphere that promotes empowerment and—in the words of the executive director— "refuses to simply make poverty bearable." While the other food programs work through volunteers coming together to distribute food, SoS is very different in that the crewmembers are placed into a new and unfamiliar situation and required to work together to make it a success. They learn the new and strange labor of urban agriculture, how the marketing of produce takes places, how to cook the foods they are growing, and how to work as a team in order to accomplish a specific task. Similar to the volunteers at CHUM, RFP, and SHARE, they have a shared task to accomplish together and use this time to come to know one another better. However, because SoS recruits unemployed individuals in need of job training their time together is more intense: they work a regular 9–5 workday together, learning the unfamiliar skill of urban agriculture.

Learning urban agriculture is an embodied process, much like learning to dance or taking part in adventure education. Conradson (2003), drawing on Nigel Thrift's work on Non-representational Theory (2007) suggests that all work in the voluntary sector is embodied and the spontaneous movements and interactions within groups are what create connections between people. To this end, crewmembers spoke of coming back home tired and achy, about the strange textures and tastes of the vegetables they were growing, and how crew leaders were always discussing with them how the plants were growing and what had to be done to further their development. While volunteers at the other programs took part in distributions and met with people similar to themselves, at SoS crewmembers were changed from being sedentary to being active, and were constantly learning new things. This means that their time together was more meaningful because they were at a vulnerable place in their life and their interactions were more intense (Smith et al. 2010).

Multiple crewmembers described their lack of relationships when they first came to SoS and how the work crew and the relationships they formed at Community Action Duluth comprised their entire community. At SoS, participants had multi-threaded connections with one another: they were living together and using joint childcare arrangements, and there was a shared culture that developed around individual self-improvement (Etzioni 2004). Phillip, a 19-year-old white man, was homeless and living in a transitional housing facility when he entered SoS and ended up living together with another crewmember he met at the program. Likewise, Debbie, an African-American single mom, moved from Chicago to Duluth and came to be friends both with the crew leaders and other crewmembers. As she described it, "Yes, I had no friends. I didn't know anybody, so anyone I know I know through [SoS]." Crewmembers would get together to cook food and also arrange for childcare together. Debbie in particular also became active in

Community Action Duluth and the range of courses and activities that they offer, such as money management courses and Big View, which create a space for low-income people to discuss public policy.

Crewmembers at SoS spoke fondly of their time working at SoS. Debbie saw herself staying involved with urban agriculture in the long-term: "I see myself doing it at my own house. Yes, I want a garden now in front of my house because it's really cool."

Dahlia, a crew leader with a background in agriculture and sustainable development (and as we saw in the previous section has strong feelings about food) witnessed her work crews forming strong relationships with one another. In her description of the groups, she talks about not only how they came to be friends with one another while working together, but also how they spent time outside of the work crews helping each other:

> The bond in the work crew is very, very deep by the end of the 12-week session, so much so that they don't always know how to not be around each other every day. They have many common barriers in their lives. They're often parents, they're often single parents and so, being able to discuss what's happening at home—how do you handle being a single parent, having adult time raising little kids at home—they bond over these things. They share resources, they watch each other's children, they go to each other's homes, and cook meals as a group. One, because it saves money, which would allow them adult time with someone who knows them, who doesn't judge them, who sees them every day, and at the end, I think there are tears. They don't know how to go on to work in a new place without this bond of someone and the relationship often sticks.

Spending time with Dahlia and the other crewmembers was interesting because they embodied so many different positions: they taught farming, ensured that crewmembers showed up on time and completed the needed tasks, but also completed work with the crewmembers and thus were embedded in multi-threaded relationships with crewmembers, a finding that many scholars of community gardening projects have noted. Community gardens are a public / private response to food insecurity because they support the communal activity of gardening as well as the private ethos of self-sufficiency through people growing their own food (Evers and Hodgson 2011). This tension between personal self-betterment and group collective action animated experiences at SoS: crewmembers had to work collectively to solve problems inherent to urban agriculture and also had to be held personally responsible, for example, to show up on time and to take part in other CAD activities.

These bonds were strong not just between crewmembers, but also between crewmembers and crew leaders. Crewmembers all described crew leaders and program leaders in personal terms, noting the intelligence of program directors who were developing (sometimes in an ad-hoc manner) an urban agriculture program. On one visit to the program, for example, the crewmembers and crew leaders were working together to figure out a way to raise and lower a set of hanging lights they were using to start seeds in an indoor greenhouse. Debbie, for example, described Paul—one of the program directors:

> I believe he's different. He's his own person, but he's really smart. He's just way more advanced than most people in his own way, so a lot of stuff, you won't know what he's talking about. You have to get him to break it down.
>
> ADAM: What are you talking about, like gardening stuff?
>
> Yes, but he uses his own big words for it, and it's like "Okay, what does that mean?" He explains stuff very good. He works with you, and whatever your situation is, he knows a way around it. He's like the most understanding boss I've ever had, and I've had a nice amount of jobs. ... We're determined because [with him] you have to be no other way.

Being a part of SoS was a way for crewmembers to not only make friends and learn new skills, but also a way for crewmembers to relate to people in the community in a new way. For example, Debbie used to come in on Saturdays to work with volunteer groups doing one-day work projects in the SoS fields. Because they volunteered for only a short period of time, the volunteers looked to Debbie to explain to them what they should do in the fields. Her role was transformed from that of an out-of-work person in need of training, to an adviser explaining to employed people how to do their job correctly. As she explained, the public

> love our work, and they all congratulate us. People ask every other day like when I walk past [the 4th street garden], "Do you want me to do some weeding for you guys?" We're like, yes. We love volunteers. Anything to help is okay.

While SoS did a good job of creating strong bonds between crewmembers and crew leaders this was a rather small community and outside volunteers were not part of this system. Paul compared the sense of community among volunteers and crewmembers:

> Typically, [crewmembers] don't know each other before the start of the season and oftentimes friendships and arrangements of mutual economic benefit form as a part of the work experience, sharing apartments, sharing child care, sharing rides. So, there's a certain amount of social capital that comes out of the trust of working with a crew for three months under sometimes trying conditions. As far as volunteers, you know, that's not so much lately that we've started having designated volunteer days. We do have some ongoing volunteers, but we tend to have the easiest time marketing what we're doing to sort of one-time church groups and corporate groups and things like that.

Middle-class volunteers at SoS served to augment the labor provided by crewmembers and were thus slotted into set work slots on the weekend that crewmembers often were not present at because they worked over the course of the week, not the weekend. This suggests that there is something especially problematic about the position of the volunteer where neoliberal subjectivities are concretized in a way that they can be challenged more by paid professionals who have more of a theory of change and empowerment.

While SoS served as a catalyst for change for those involved in the programs, participants and volunteers at SHARE, RFP, and CHUM all expressed an interest in urban agriculture and many had their own backyard or community gardens. Most downplayed the size and quality of their gardens because of limitations posed by health conditions or lack of easy access to space for gardening, but still expressed an interest and excitement in the process of urban agriculture. Bernadette, an RFP participant, responded to a question about her experience with urban agriculture noting, "Well we started one. We've got one of those four-foot, three-foot garden-in-a-box, and it's a raised garden." She grows lettuce, onions, and peppers in her home garden, but explained that her father (who really has the time to engage in gardening) cannot help because of his arthritis and fibromyalgia.

Urban agriculture was a food program that participants in all of the programs saw value in not because of a well-articulated desire to transform the food system, but because it seemed like a simple and hands-on way to improve their own food quality and life. This finding is consistent with Evers and Hodgson's (2011) work on urban gardening in Perth, Australia where they found that community gardens improve food security both as spaces of *direct* food production and as *indirect* spaces of education and community empowerment where people learn about food production, get-to-know their neighbors, and discover new kinds of eating. For participants in all programs, gardening was an unassuming and enjoyable task that some had been involved in in the past but now were not; however, they still had fond memories of gardening. Sarah, at 49-year-old Native American CHUM participant, for example, remembered growing up on a farm where there was "an acre full of vegetables, strawberries, we had a garden of everything…" and saw gardening in Duluth as a way of getting back to some of that simple access to vegetables. While survey respondents did not purchase organic or natural foods in their normal shopping, they saw gardening as a way of accessing foods that were clean and wholesome. Katherine, an RFP participant, supported gardening just "because it's a chance for people to get fresh vegetables that don't have a lot of pesticides sprayed on it."

While the organizers of SoS necessarily approached the urban agriculture process as an organized and logistically complex operation, participants from all of the programs spoke about backyard gardens that were not formalized but were productive in their own way. Alyssa from RFP practiced urban agriculture, but described herself as not really having a garden, just beds in terraces where she couldn't quite remember what was planted: "I don't really have a garden. I had the tomato plants along the wall of that side of my house and then like a terrace, because I live on a hill. There's three terraces and the middle one is where he planted the corn and carrots, and I think he did onions. I don't remember." Similarly, Hettie from RFP described her inability to garden because of problems walking without a cane, but expressed an interest in still being a part of this turn towards urban gardening:

I would like to, I don't know if I could. There are days when I crave, days when I cannot walk without a cane, I love to be in that, and I love the feeling of growing something and eating what we are growing, and I kind of vaguely knew there was [a garden] here, but unfortunately priorities change when your life changes drastically because I used to volunteer a lot too; I don't anymore but it would be something I would be interested in getting information about for sure.

The significant road blocks to permanent involvement in urban agriculture meant that people who were growing things didn't always view themselves as being part of an agricultural process. Jodi, a 51-year-old homeless white CHUM participant, for example, didn't imagine herself doing urban agriculture and initially said she wasn't growing anything, but later admitted to growing herbs in her garden, which was a "quick" and "easy" way to get access to fresh flavors:

Myself, no. I have artificial knees too so I know that it would be a difficult thing for me to do [urban agriculture]. I remember being a kid and my grandpa having a garden, and it was so fun to eat like the first peas and carrots and stuff. Herbs I often do in my apartment, yes. That is easy to do.

The accomplishments at SoS creating community questions the traditional model of surplus food distribution centered on volunteers providing for an in-need population. By placing crewmembers at the heart of the operation, volunteers are shunted off to an auxiliary role of accomplishing labor-intensive projects under the direction of crewmembers. This allows crewmembers to be recast as knowledgeable directors of the labor of others instead of individuals in need of help. Crewmembers are then able to spend the majority of their time working closely with one another building community and learning the new skill of urban agriculture. The project also moves away from addressing food insecurity through surplus redistribution. Instead, in growing their own food, food becomes a product a labor created through the hard work of SoS crewmembers.

The downside of the program is that it operates on a small-scale basis. As Rebecca, a director of the program, noted SoS is "super small and I have no illusions about that it is changing the food system in any way. No, I think that needs to be dealt with, but that's not what we are doing." The program works through creating small-scale personal relationships between crew leaders and individuals and thus in order for it to grow and effect more people, it would need to fundamentally change its design and operation (for an interesting analysis of "scaling up" see Friedmann 2007).

One way to read the difference between SoS and the other distributive programs is to argue that because urban agriculture connects people to the land and questions the commodification of food, it is a much more radical intervention into the US food system and therefore better able to transform lives and build community. French Marxist social critic Guy Debord, building on the work of the Frankfurt School of critical theory, argues in *The Society of Spectacle* that

"In societies where modern conditions of production prevail, all of life presents itself as a mass accumulation of spectacles. Everything that was directly lived has moved away into representation (1, 1983[1967])." Thus the commodification, overproduction, and surplus of foodstuff in the US has changed food from a concrete object imbued with meaning to a ubiquitous object devoid of specificity and significance. To Debord, true community cannot be formed in conditions of mass production because "the spectacle is not a collection of images, but a social relation among people mediated by images (1983[1967], 4)." Hence, in the US—where less than one percent of the population is involved in the actual production of food, and food production and marketing have become an all-consuming spectacle—meaningful relationships cannot be formed simply through changing the venue and the price of these commodities. While SoS revolves around eliminating the distinction between producer and consumer, CHUM, SHARE, and RFP distribute just the types of spectacles that Debord critiques: prepackaged mass-produced food intended for broad distribution. In fact, were it not for this overproduction, CHUM and RFP would not be able to have food available for distribution. Aspects of the Good Movement, such as farmers markets, are often aimed at defetishizing food and making visible the connections that exist between farmers and consumers (Alkon and McCullen, 2011). However, farmers markets have been critiqued for their inherent whiteness, neoliberalism, and lack of focus on the food access needs of low-income consumers (Alkon and McCullen 2011; Slocum 2008; Guthman 2008).

Debord's critique of community in modern society is echoed in the writings of many social critics as well as members of the Good Movement. One of the foundations of The Catholic Worker Movement as expressed by Peter Maurin (1949) was the development of rural cooperatives where the unemployed could learn job skills and escape the alienation of the city. Further, Ferdinand Tönnies in *Gemeinschaft and Gesellschaft* (1957[1912]) unfavorably compared the multi-threaded connections of traditional rural communities to the rule-bound set relationships that govern life in industrial society. Likewise, Peter Kropotkin (1906) argues in *The Conquest of Bread* that a relationship to the land is essential to the development of a just and compassionate society. He argues, "The large towns, as well as the villages, must undertake to till the soil. We must return to what biology calls 'the integration of functions'—after the division of labour, the taking up of it as a whole—this is the course followed throughout Nature (118)."

While it is tempting to argue that SoS was able to break through the spectacle and create community based on a connection to the land, the barriers to community formation at the distributive food programs were the result of their problematic distinction between program participant and volunteer, strict sets of guidelines that were enforced on participants, and overriding focus on the logistics of providing food to people in an efficient manner. At SoS, volunteers worked with crewmembers to grow and market food, solve problems inherent to the small-scale production of food (for example, how do you hang grow lights in an old school building?), and this served to bring people together. The task was less important than the communal acts of problem solving. Similar to the leadership teams at RFP and SHARE, the workers at SoS were involved in a hands-on project that demanded trust and accountability to work, and this led to strengthened connections and community.

Creating Community and Providing Access to Food

Forming community within the vibrant space of the voluntary sector is a complex project. With the rise of neoliberalism and the devolution of poor support programs to the non-profit sector, programs that distribute food provide a vital component of the social safety net and must create an environment that encourages volunteers to return week after week. Even though there is ample contact between participants and volunteers, each group is somewhat distinct from the other, and the class and race divisions that define US culture in general are reified within the shadow state. At CHUM, RFP, and—to a lesser extent—SHARE, volunteers found themselves united with other people "like them" and formed an affinity group of people interested in serving the community. Not only were volunteers drawn from the same group of giving and civic-minded individuals, but the ritualistic process of getting together once a month to unload trucks and parcel out food became an important part of the "particular culture" (Etzioni 2004) that volunteers created with one another and enjoyed being a part of. The good spirits of volunteers contrasted markedly with low levels of engagement that participants brought to the programs. They saw themselves as having little in common with other program participants, and as such were somewhat passive recipients of food instead of empowered participants in the programs. They had no common task or mutually supporting culture with which they could feel more attachment, and instead used the programs to solve the specific problem of food access.

The cultural disconnects between program volunteers and participants were echoed in how each group related to food. For volunteers, the preparation and procurement of food cemented relationships and was used to express their chosen relationship to the current food regime. In contrast, participants were much more utilitarian but no less purposeful in their understanding of food. The diverse ways in which they utilized the resources available to them were not incorporated into the structures of the programs, but instead existed as a distinct set of individual knowledge. Community emerged from the case studies as a negative force at CHUM and RFP, and to a lesser extent at SHARE, in that it underscored the class and race differences that separated volunteers from clients.

These findings suggest that the organizational structure of the program played a strong role in how a sense of community was formed among participants and volunteers. The inherent power dynamic in serving others, be it through a leadership team at RFP and SHARE, or the traditional volunteer labor of CHUM, worked against community in these projects. In light of the differences in how volunteers and participants thought about food, SoS created a noteworthy space where the downtime inherent in production and shared meals allowed crew members to eat and taste new foods. Urban agriculture was the platform upon which these relationships developed. Interestingly, many participants at all of the programs saw urban agriculture as a project they would like to be involved in, indicating the strong pull of self-provisioning and spending time working with the earth. As will be explored further in the next chapter, the ways in which program participants carefully structured their food supply and saw value in urban agriculture are resonant with the community economy literature that values forms of development outside of the dominant capitalocentric models.

SoS was not a space that lacked tension or intentionality about the connections between food and health; in fact, staff members openly discussed the idea of offering yoga classes as a way of addressing the mental stress or allostatic load of poverty (Schulz et al. 2012). In this sense, the ethos of the project demanded both a recognition of the ideal of healthy eating, and the reality of the present broken food system. Paul from SoS described how he walked this tension:

> It's an invitation to try out [new foods], say you hadn't really learned how to cook whole foods, you know, trashing people's food culture is not a productive way to get them to try something that is probably more nutritionally beneficial. The great Catholic worker, "White Sugar Debates" where people in an intentional community that provide hospitality for people that are homeless debate about whether or not it's acceptable to have white sugar in their house for use with coffee, because that's what guests expect, because their values say that's not an acceptable food and that contrasts against the value of hospitality in terms of providing what people expect and what make people feel at home. I tend to come down probably on the permitting white sugar and inviting people to try out turbinado [sugar] as a side thing.

By offering this invitation and providing a space for this offer to be explored, SoS affected not only how people relate to one another, but also how they ate. However, this was only possible because of the intimate scale of the operation: their intervention does not address the societal problem of hunger at the scale that RFP, SHARE, and CHUM have designed their intervention.

In the next chapter, I move away from the concept of community and examine the ways in which alternative economies are formed within each of the food access organizations. In contrast to the dominant narrative of neoliberal urbanism that focuses on revanchist urban forms and the limiting of public space (Jessop and Sum 2000; Lederman 2014; McCann 2011), each of the case studies worked strenuously to serve others, and the projects were enacted by individuals who were motivated by a deep sense of service to the marginalized. Beyond simply acting on their ideals, they set up institutions with boards of directors that committed to making decisions in light of ethical considerations.

Notes

1 Peter Maurin. *A Question and an Answer on Catholic Labor Guilds.* "Catholic Radicalism," 1949. Page 20.
2 © 1995. John McKnight. *The Careless Society.* page 108. Reprinted by permission of Basic Books, a member of the Perseus Books Group.

4 The Emerging Alternative Economies in Community-based Food Programs

The Fallacy of Savings

When people save money, they invest that money. Money invested increases production. Increased production brings a surplus in production. A surplus in production brings unemployment. Unemployment brings a slump in business. A slump in business brings more unemployment. More unemployment brings a depression. A depression brings more depression. More depression brings red agitation. Red agitation brings a red revolution.[1]

Poverty, we have said elsewhere, was the primary cause of wealth. It was poverty that created the first capitalist; because, before accumulating "surplus value," of which we hear so much, men had to be sufficiently destitute to consent to sell their labour, so as not to die of hunger. It was poverty that made capitalists.[2]

Arriving at Ruby's Food Pantry on a freezing cold winter afternoon is like stepping out of your everyday life and into a giant performance of generosity that has been enacted so many times the actors perform their roles like clockwork: a volunteer dressed in snow pants, goggles, and an enormous winter coat has icicles hanging off his beard as he directs traffic in the parking lot; a forklift brings pallets of food from a semitrailer into a brightly lit gymnasium for distribution; official-looking women sit at a table in the entranceway in red Ruby's Pantry T-shirts make sure new volunteers sign in and put on nametags. There are so many projects going on that newcomers enter the gymnasium and just stand at the door befuddled about where to go next: one group is digging through a box of potatoes the size of a small car; another group is chipping away at a similarly sized box of frozen pre-cooked scrambled eggs; kids are throwing empty cardboard boxes out of a window and joyfully watching them land on the overflowing dumpster below; and volunteers are carefully arranging items along tables, waiting for the crowds to arrive. No one is being paid—everyone is here to be part of this project because they believe in the mission of giving back to others. And RFP is not unique, each of the food access organizations in this study is involved in a project that puts into action an alternative understanding of how decisions about resources should be made. In their one-on-one interaction with people experiencing food insecurity, they validate the pain and suffering that hungry people experience. In the act of providing food to the needy, they transform this ubiquitous commodity into a symbol of god's love and the community's care for the hungry. And, and in their ideology as organizations, they make decisions according to ethical considerations and assert a fundamentally different way of organizing society and welfare than the neoliberal norm. While programs such as Food Not Bombs (Heynen 2009) and the collectively managed Vegetables Unplugged CSA (Wilson 2013) were created by ideologically motivated activists committed to creating "autonomous spaces" (Chatterton 2005) where interpretations of a "new political imaginary" (Gibson-Graham 2006) could be put into place, the programs in this study place food distribution and care for the individual first and are not involved in a specific overarching critique of capitalism. However, in their actions and program design there is a commitment to making decisions according to deeply felt moral ideals about how resources should be allocated in society and how institutions should treat individuals. While, as will be explored in Chapter 5, these visions were not completely put into place, they illustrate the potential for the shadow state to operate as an alternative economy.

In the first chapter of their 2013 guidebook on how to create community economies Gibson-Graham et al. propose the analogy that an economy is like a community garden: it is fueled by the natural sunlight and our life-producing ecosystem, volunteer laborers provide the work to make it productive, and after producing vegetables for consumption by the gardeners there is "surplus production" to give away to the local food shelf and sell at a local farmers market. The proceeds from these sales are then reinvested back into the garden through a process mediated by the gardeners who run the plot. They sum up their analogy writing:

> Economies are basically no different from this garden—each economy reflects decisions around how to care for and share a commons, what to produce for survival, how to encounter others in the process of surviving well together, how much surplus to produce, how to distribute it, and how to invest it for the future. These decisions are made under varying conditions of plenty and scarcity.
> (xvii)

As opposed to simply reading community-based food access programs as institutions of the shadow state, we can also read them as emerging community economy projects: places where decisions are made according to ethical considerations, surplus is distributed based on democratic decision-making, and positionalities of food-insecure individuals as disempowered and marginalized are challenged. But this is a tricky position: is selling surplus production at a farmers market ethical if that farmers market embodies class and racial prejudices (Alkon and McCullen 2011)? And what are the moral problems with distributing surplus through the food shelf that has been critiqued for promoting clientelist subjectivities among its clientele (Poppendieck 1999)?

This chapter explores how diverse economies are enacted or are emerging within each of the community-based food programs. Each of the community-based food programs in this study engage in a set of practices that are broadly consistent with the ideals of alternative economies: they make decisions in light of ethical considerations and their program design and operation, ontologically reframe subjectivities, and challenge traditional neoliberal economic concepts. At each food program, I describe their overall mission and food project and explore how their organizational focus and the changes that they have made over time reflect their commitment to an ethical ideology. In sum, they challenge neoliberal norms by articulating a different relationship between people and food, and creating institutions that make decisions within the boundaries of self-imposed ethical constructs. Next, I focus more intensely on the provisioning strategies of program participants and explore how these "clients" can be theorized as "agents" fully engaged with diverse economies through reliance on friends and families and their engagement in non-traditional food gathering practices such as gardening and hunting. Last, I explore the many ways that program participants would like to see the organizations that they are involved in grow and expand into other aspects of service. I use these possibilities to analyze the concept of transforming the shadow state *from within*, through those individuals dependent on its actions more actively involving themselves in its governance.

Creating a Faith-based Economy of Care at CHUM

CHUM is a coalition of 40 congregations that work together to provide services to the poorest and most marginalized communities in Duluth. In order to engage in this mission, they have a drop-in center that provides services to Duluth's homeless population, a homeless shelter, a transitional housing facility for families, job-training programs, and a community organizer who works on issues facing the low- and no-income community, as well as state-level lobbying. Member congregations are woven into the fabric of CHUM's services: each evening a different congregation serves dinner at the drop-in center; beds and quilts at their new permanent supportive housing facility were handmade by members of the CHUM community, and when the shelter was overflowing a member congregation opened up its basement to operate as an overflow space. Therefore, while CHUM staff performs much of the labor of the organization, it is augmented by the work of member congregations. CHUM's name has gone through a number of iterations, starting as "Central Hillside United Ministry," although they are now known only as CHUM in order to recognize their non-Christian member congregations. Although mainly Christian, they also have Muslim, Jewish, and Unitarian member congregations. Because the organization accepts a wide variety of state-funding sources and embraces a policy of unconditional embrace for people struggling, outward symbols of religious identification are in large part absent from the establishment, and religious identity is much more present on the board of directors and in the motivations of individual volunteers and employees. For example, board meetings begin with a prayer, a full-time congregational outreach staff member works to involve member congregations in their work, and much of the staff is motivated by religious ideals.

The food shelf began in 1979 as part of the Duluth Food and Nutrition Council and became a part of CHUM in 1981. CHUM's model of food distribution has evolved over time, but these changes have been slight. Around ten years after they opened, they switched from a model where participants could only use the food shelf if they had been referred by a social worker, to the current model where participants come of their own volition but are encouraged to use the shelf only once a month. In response to rising hunger in different parts of the city, CHUM opened up two other locations: a West Duluth branch started in 1982 and the East Duluth in 2013. These sites are smaller than the main location where this research took place.

While neoliberalism seeks to make decisions in light of economic rationale, CHUM's executive director writes in their annual report that they are guided by the questions "given this unique and God-given person in front of me at this moment, how can I respond with the most love, the most generosity, the most compassion, and the greatest mercy? How can our work together help heal the broken world and bend the moral arc of the universe a little bit more toward justice?" (2013). Williams et al. (2012) write, while "roll-out" neoliberalism has embraced faith-based organizations as "little platoons" of the state, within these organizations "it has been demonstrated that the incorporation of faith-motivated activity can enable subtle but significant shifts in moral and ethical politics *from within*" (1489). Thus, the religious ideals of the organization shape how they serve the poor, and, as a parastatal organization, shape the "locally contingent" forces which determine how state policy is enacted at the local level (Brenner and Theodore 2002). CHUM challenges neoliberal social policy and enacts their vision of justice through three specific practices: community organizing, challenging neoliberal subjectivities, and addressing the needs of the poor in a comprehensive and non-piecemeal way.

CHUM uses its size and organizational capacity to put pressure on the state to respond to the needs of low-income citizens at a variety of different spatial scales. CHUM is unique among corporatist voluntary associations in that they employ a full-time community organizer whose job is to advocate for the needs of the low- and no-income community in Duluth. This project echoes Gibson-Graham et al.'s (2013) assertion that "*in a community economy we take ethical action by acknowledging how our survival is connected with that of others*" (39, italics in the original). A committee that includes homeless and low-income individuals, as well as other members of the CHUM community, oversees the position and helps to set the organizing agenda, and illustrates the project's commitment to democratic governance. The organizer worked with state and local housing agencies to organize a tenants union in a local Single Room Occupancy (SRO) facility that provided essential housing for Duluth's low-income population, but was in very poor repair and under terrible management. The community organizer has become active in local efforts to address food insecurity through the organization Fair Food Access Lincoln Park (FFALP). This group is using traditional community organizing efforts, such as door knocking and the development of community leaders, to increase community capacity to push for better bus access to local grocery stores as well as putting pressure on local convenience stores to sell healthier foods in the neighborhood of Lincoln Park, which has been identified as a food desert (Pine and Bennett 2013). By working collectively to improve neighborhood living conditions, FFALP has reclaimed community space through the creation of gardens and improved neighborhood empowerment by pushing government and private businesses for better treatment.

This activism "jumps scales" (Smith 1992) and addresses issues beyond the confines of Duluth. CHUM partners with the Joint Religious Legislative Coalition (JRLC) and Take Action MN to press for the needs of low-income people at the state capital in Saint Paul, often sending busses of pastors and citizens to lobby in the state capital. As a part of this project, they have recruited and trained low and no-income individuals to testify in hearing at the state capital around healthcare reform in Minnesota. This project of political advocacy is directly related to political empowerment, seen by Watts and Bohle (1993) as a key part of hunger. They write "political exclusion from the process of entitlement distribution and enforcement is therefore central" (120) to the development spaces of vulnerability. Importantly, this organizing work is directly linked to their religiously guided ethical project of helping those in need. As part of their support for legislative solutions to problems facing their service population CHUM hosts a legislative send-off, where they invite members of the CHUM community to pray for the region's legislators and encourage them to keep the beliefs and needs of marginalized people in their heart. The 2013 send-off concluded with the following prayer, which draws connections between faith and public policy:

> Wonderful God, you have challenged us to work together to craft a society of justice and peace. You have placed us in a world of beauty and of sorrow. Pour out your spirit on these women and men as they go to serve your people. Give them vision, compassion, dedication, courage and love. Remind them of their neighbors here in the Northland—those who struggle, those who reflect your love in their work, those who hope for a world where it is easier to be good. Stand by them when they seek to speak truth; give them strength when they work for the poor. And give them time to rest in your peace when their work is done. Amen.

A further example of this commitment is that while many of the programs offered by CHUM are funded through grants and payments from CDBG and the city, the community organizer position is funded through donations from member congregations and strategic use of contracts which allow the position to be paid for to provide services while simultaneously organizing protests and collective actions. This process is consistent with "rule bending" (Lipsky 1980) wherein the rules of organizations are unofficially altered by sympathetic staff members in order to better serve the needs of clients and substitute their ideology over the official ideology of the organization (see also Maron 2012).

CHUM's actions that alter the political subjectivities of low- and no-income people is grounded in their familiarity with the day-to-day struggles, needs, and successes of this population. For example, during the 2014 and 2015 election cycles they partnered with other organizations—including CAD—to host a "reverse forum" on poverty, with low-income people asking questions and making specific policy demands of candidates for political office. Under this format, the needs of the low-income population for assistance from the state were made important and aired in a public setting, attended by news outlets, and politicians were given merely one minute to respond. The project is furthered through an annual Christmas Eve Vigil hosted on the steps of City Hall to remember the lives of members of the drop-in community who have died over the last year. The vigil is a heartbreaking event attended by clergy and members of the extended CHUM community, including many homeless individuals as well as utilizers of CHUM services. At the 2013 vigil, the officiating pastor reimagined the Christmas story of Joseph and Mary looking for a place to stay in Bethlehem as a story of homelessness, beginning by saying that "the society in which Jesus lived was characterized by what we today would call, 'radical hospitality' where a person could knock on any door and expect a safe—if not glamorous—place to sleep." He told the story noting the role that community plays in creating this type of welcoming for those in need, concluding:

> I think that this story tells us that the *essence* of home consists, not merely in the physical structures, important as they are. The essence of home consists in the network of real and healthy relationships that surround us and shelter us and include us, that give us a place in the world, however humble. To have a home is to belong, in the most fundamental sense.
>
> The true and terrible homelessness is "not belonging." It is failing to recognize our own neighbors. And this disconnection, this lack of community, is something we may suffer, even if we are warm and well housed.

Here, we see the creation of a religious ethos of care that unites volunteers and program participants under an overarching understanding of what "community" means in a fractured and imperfect city. This argument crystalizes many of the calls to activism championed by Catholic Worker Movement (Day 1963), which advocated the opening of houses of radical hospitality completely outside of the state welfare system in order to provide whatever services were needed by those suffering from unemployment and homelessness.

A final example of CHUM's use of their position in the shadow state to advocate from a faith-based perspective is the construction of the Steve O'Neil Apartments, 44 permanent supportive housing units for families with children who have a history of recurrent or long-term homelessness. The goal of the homes is to echo some of the work of Community Action Duluth in that they will serve as an all-encompassing place of support offering children's programming, employment assistance, and shared community space. As the blessing for the apartments ended, "Remind them frequently, O God, that they are ours and we are theirs. And we pray for them."

The limitations of CHUM's activism and programmatic activity are typical of organizations that operate within the shadow state: because of their necessity to receive both money from the state and member congregations, their actions must receive broad-based support in order to move forward. Rose (1999) argues that community can be used as a form of discipline, and we can see this constricting CHUM's efforts to fight for the marginalized. For example, their community organizer has traditionally been careful about too much outright antagonism against local landlords renting to low-income Duluthians because of the need to place homeless CHUM clients *into those very same units*. Similarly, as a coalition of 40 congregations that include Catholic as well as mainline protestant denominations, their need to receive unanimous support for efforts often means not advocating for political divisive issues, such as same-sex marriage, which many consider to be part of the bundle of human rights that the organization stands for. Further, as explored in Chapter 3, their enthusiastic embrace of faith-based activism is somewhat separate from their contact with clients. Given the religious diversity of the city (both interdenominational and interreligious) their strategic use of religion is savvy, but often means that the specific belief systems that bring volunteers and congregations to activism are not addressed with clients.

CHUM is a comprehensive faith-based social service agency within which the food shelf plays a small and vital role. The organization publicly embraces an activist form of faith within which service to the poor and care for the marginalized is a sacred and necessary part of their way of being in the world, and where activism for their rights is part of the organization's structure. However, while volunteers and board members express this openly, services are offered to individuals in a secular setting, which works against the creation of a common language among participants and volunteers about the role of faith in social transformation. Of course, embracing this faith more openly would mean alienating participants, disturbing the coalition of churches that support CHUM, and the loss of federal funds. Participants speak highly of the services they receive from CHUM but indicated that they would like to see other services offered by the food shelf such as prepared foods and more meat.

Distributing God's Abundance to All Comers at Ruby's Food Pantry

RFP grew out of the work Lyn Sahr, the founding pastor of Pine City Community Church in Pine City, Minnesota and his multiple evangelical projects. In 1997 he began a daily e-mail and web-based spiritual "thought of the day" service known as *ChurchMouse Chronicles,* which came to be seen by thousands of people (2014). In 2002, this project grew into a separate program known as Home and Away Ministries whose mission is to "to support pastors and churches for the common cause of Christ, to develop cross-cultural relationships between churches in America and other countries and cultures and use corporate surplus to help fight poverty, hunger and disease at home and away from home" (2009). The group is an evangelical Christian non-profit, with a clear statement of faith outlining their religious beliefs and a board of directors drawn from the local Minnesota and Wisconsin region. The "away" portion of the ministry involves leading trips to Monterrey, Mexico where they bring truckloads of school supplies and toys, lead conferences, and build community with other evangelical churches.

During a search for donations to bring to Monterrey, the group was surprised to hear from potential givers with a desire to donate things other than the goods that were requested. For example, food donations were offered along with toy donations, and school supplies were offered along with somewhat random donations such as tires and copy machines. The disbursement of these donations to those in need came to be the "home" portion of the ministry, consisting in large part of RFP. These donations were offered to Home and Away Ministries as tax write-offs—a convenient and religiously grounded way to dispose of unneeded items—and RFP became adept at meeting the needs of companies looking to rid themselves of goods and providing those goods to those in need. Thus, RFP began to serve the role of distributing the surplus created by the industrial food system and reframing this abundance as a gift from god to be distributed by religiously committed individuals. For example, RFP sends trucks cross-country to receive goods and because they are small and flexible they are able to work around the needs of corporations instead of asking corporations to work around their schedule. In this sense, distribution of surplus has always been at the heart of the RFP project.

RFP has been around for three years and has at its core the neoliberal concept that the $15 fee for a portion of food is what makes recipients eligible for their largesse. Three aspects of RFP's program illustrate the ways in which they operate as a community economy: their move toward comprehensiveness and responsiveness to the voices of participants, development of a community controlled slush fund, and development of an ecumenical coordinating committee. While at the first distributions there was nothing for participants to do as they waited for their food, they have now invited a local healthcare access agency to distribute literature, and a subsidized dentistry van also has been parking at RFP during the distribution. These new resources do not mean that RFP has radically altered its relationship to the state: there is still no screening process or help signing up for SNAP benefits as CHUM and SoS engage in as RFP define themselves as a program catering to the needs of individuals who earn too much to qualify for state-offered benefits but still find themselves in need. However, in offering these other services RFP has started to transition from solely a food access program to a comprehensive community-based institution guided by an ethical commitment to social change.

In making this change, they are also responding to the wishes of their participants and illustrating that although there is not a formal space for participants to participate in governance their voices are being listened to. Gibson-Graham argue for democratic decision-making in community economy projects, and under the RFP format the coordinating committee has decided to expand their offering based on their internal discussions, illustrating a representative if not democratic process. Interviews with RFP participants indicate that many would like to see the program evolve into a project that is easier for them to access, offers a greater assortment of goods, and where the distribution site itself is more family-friendly. Interviewees talked about their desire for more distribution sites with more flexible hours so that the logistics of making it to RFP each month were easier. For others, food quality was an issue and they wanted more fresh produce and healthier food as opposed to simply the overstock food that RFP specializes in. Shannon described the ways in which RFP could expand:

> There's a lot of people and some of them are interacting in their little niches like talking to their neighbor or whoever they came with, but I don't see a sense of community. I think it would be really super interesting and wonderful to figure out a way to create a sense of community so that when you came there and you had to sit for an hour, you wouldn't sit in [the sanctuary] miserable with your eyes cast down, but you were actually engaged. I don't know what that would look like. I don't know if that would be an aerobics class, or a yoga class or a sing-along, I don't know what that would be. But I would like to see a way for there to be a sense of community and joyfulness rather than the sense of what I see is a sense of dreariness. And it's not the fault of RFP. It's just kind of the nature of admitting that I might need to take advantage of [a program like] this.

Because RFP is limited by the constraints of all voluntary sector projects these changes may not occur, however, utilizing the views of program participants to reimagine how care is given to others illustrates an exciting space of possibility for the program.

The biggest change for the local coordinating committee is that RFP has developed a small "slush fund" by saving a small portion of the $15 distribution fee to invest in things needed for the event. They have purchased box cutters and plastic bags to help with the distribution day itself, as well as dollies to assist in moving heavy boxes of food for clients. In alternative economy terms, this "surplus" has become a community asset for the coordinating committee to distribute as they choose. As Gibson-Graham et al. (2013) note, surplus "is one thing we have to work with to create new worlds" (65) and they advocate a process of collective discussion and open analysis of the costs and benefits associated with how to disperse this product created by collective work. RFP is following just this process through branching out and using those funds to invest in other community food projects such as the local food bank and household necessities. Nathan described the transition of RFP in the local community as they have come to disperse this surplus:

> Our goal when we first started the program was that by the time we are a year old we would be able to take those funds and go and make purchases like bar soap, laundry soap, dish soap, toilet paper, and be able to offer something like that to each one of the patrons. That is something different, supplementing what we are doing. We have also had discussions to how to better distribute funds to organizations that require them and I think we are getting through that. So that is part of our work, "What can we do to help in other areas of the community besides giving food to the patrons?"

This change signals a transition of the local coordinating committee from an ad hoc group of religiously motivated food activists to an organized group with a developing ideology and the power to put those desires into action. It also illustrates how the distribution of this surplus can be done through an evangelically based analysis of the needs of the participants in RFP.

A last example of RFP acting as a community economy is the ecumenical nature of their coordinating committee, and the way that they have worked to bridge differences in theology in order to come together in the spirit of community change. Here, you see the coordinating committee acting according to what Gibson-Graham (2006) would label "practical ethics" or "weak theory" in that their vision of food access is more important than their attachment to a pre-conceived notion of how to manage and theorize their project. While CHUM unites disparate congregations, they share a similar commitment to social justice. In contrast, RFP combines Evangelical Christian churches and a Unitarian congregation under a commitment to food access. This has been done through negotiations on the committee around how to utilize prayer and proselytization.

The RFP distribution takes place at a church and people wait in the sanctuary for the opportunity to go downstairs into the gymnasium to collect their food. There is not an explicit mention of the religious nature of the organization to the crowds waiting in the sanctuary, however, there is an opening prayer for volunteers, and the committee has worked to resolve differences concerning evangelization. At an early organizational meeting, the committee had a spirited discussion about the role of RFP as an evangelizing opportunity when committee members reported back from a conference of RFP participating churches. At the conference, churches were advised to be open and friendly and to ask to pray with people who seemed in pain. Some thought it was a good idea, while others though this would alienate potential RFP users, and put organizers in a difficult position.

This concept divided the local interfaith coordinating committee, many of whom argued it was better for volunteers to simply be a welcoming presence in people's lives as opposed to offering specific invitations to pray. A similar debate emerged around the idea to play worship music in the sanctuary as participants waited for their food to arrive. In the end it was decided that if people wanted to pray volunteers should do this, but they should not make the first offer. Similarly, worship music is not played in the sanctuary because it added turmoil to an already busy setting. Nathan, an RFP volunteer who was very active in the administration of the program, outlined his understanding of RFP noting that this could be a way to get people into a church who perhaps were not regular church attenders before using the program:

> By having [the distribution] at a church we might be getting people through the door that they have never been in—does that mean that they are going to have to come back? We hope so but that does not have to, at least they have stepped in and they know that it is a faith-based program and somehow, I think, I see it as if they are accepting the fact that there is a god and that he is caring and that he is caring for them through us. [If] they recognize it or not, doesn't matter.

Faith, however, was not a prevalent theme in the voices of program participants, instead, their commonality with organizers was in their excitement to receive the abundance that was offered by the program and finding out that there were others in civic society interested in helping them. Tony, a 61-year-old white veteran described evangelism as "one beggar telling another beggar where he found a loaf of bread" and described religion at RFP in these terms:

> Most of them volunteer because ... How to put this? There are a lot of people who come there who have probably no spiritual background, but it helps them financially. But the basis for people volunteering, especially through church organizations, is that you can go out and do a street ministry on the street, take your bible, walk down to the red light district and tell everybody, "Be of good cheer. God loves you." And they're going to tell you which side of the bus to get off, because that's not where they're at. It's evangelism, which is basically one beggar telling another beggar where he found a loaf of bread, right? So if they're hungry, then you tell them where they can get a meal, or maybe give them one. If they're cold, you get them a coat, or show them where they can get one. They may ask you why you did this. And you can tell them. They may still tell you to get off the bus. But first you meet the needs, and then you can tell them why you did it.

The emerging community economy of RFP is built on the idea that individuals working outside of government can use surplus production to support the hungry. Their founder Lyn Sahr writes "When we first started Ruby's Pantry, it was with the idea of helping people who were trying to make it on their own but just couldn't quite make ends meet. Yes, a hand up if you will. When we started we had no funding of any kind and we still receive no federal or state funding. But somehow I believed that God was in this and we moved forward believing in His divine guidance" (2014). At the local level this project has brought people of different religious traditions together who have utilized "practical ethics" in order to bridge these differences and in order to provide food to those with the resources to pay a $15 fee.

SHARE's Commitment to Empowerment

Self-help and Resource Exchange (SHARE) has as their central theory the idea that collective action can be used to effectively address the food needs of community members. Of the different case studies, they engage in the most concrete way with the community economies project as their goal is to instill in communities collective and community control over the economic process of food allocation. In this sense, they engage in the process of empowerment by creating leadership groups at each distribution site and put the community in control of the allocation of surplus. Although limited to food, the SHARE project appropriates the surplus that is typically expropriated by business and instead uses volunteer labor to lower food costs.

SHARE began in San Diego in 1982, during the recession, as a way for food activists to provide affordable food to low-income and unemployed individuals, many of whom were newly impoverished and not participants in other government or civil society food-support programs. The founding members of SHARE had a diverse set of backgrounds including UNICEF, private industry, The Hunger Project, The Teamsters, the Catholic church and the Ecumenical Coalition of Concerned Americans—a group that was doing a distribution in a manner similar to what has become the SHARE model. The leadership team formed and designed the program as a method of food distribution distinct from the food shelf system and the whole foods and co-op movement, which also started around this time. As one founder described this ideology, "If you go to a food shelf you might feel embarrassed or ashamed because the food shelf is based on scarcity, and SHARE is based on 'aren't I smart because I'm stretching my food dollar!'" Participants at the first SHARE distribution always paid for their food and the group did not distribute whole foods or surplus food. Instead, prices were kept low through bulk ordering and the strength of volunteer labor. The food distributed was a mix of meat, fresh fruits and vegetables from the fields around Sand Diego, and other grocery items. SHARE does not usually start new distribution sites, instead they work under a model similar to Saul Alinsky, who saw urban communities as organized, just not organized to mobilize for their own self-benefit (1971). Distinct from RFP, which embraces payment for services through the rubric of individual self-sufficiency, there has always been a focus on empowerment through control of the economic process of purchasing food and an embrace of the communitarian ideal of working together to solve group problems. The opening founders of SHARE, for example drew on the Freirean concept of conscientization (1970) as they explored how the unequal process of food production and how their interjection into this process could be designed.

The complicated idea of community empowerment is theorized by John Friedmann (1992) in his work on alternative development. Friedmann defines social empowerment as composing three aspects: household, political, and psychological. *Household* empowerment addresses the presence of the necessarily skills and knowledge to engage in the necessary acts of household reproduction. *Political* is the ability to have their voice heard in collective decision-making processes, and *psychological* empowerment is the individual agency and self-confidence to envision themselves occupying a different role in society. This conception of empowerment dovetails neatly with Watts and Bohle's (1993) theorization of spaces of vulnerability, as volunteers gained the psychological empowerment to join coordinating committees and be involved with food distribution process, and organizers worked together to learn about the economy of food production and distribution. As Gibson-Graham (2005) write about this project of collective learning: "what we came to recognize is that desire stirs and is activated in embodied interactions and settings in which power circulates unevenly and yet productively across the many different registers of being" (133). In this sense, the coproduction of knowledge in the creation of community projects both empowers individuals and unites communities.

SHARE works with existing community organizations, such as churches, senior centers, youth groups, and unions to develop teams that will create their own distribution site. SHARE's role is to administer the warehouse and select the food to be made available, while the local team unloads the truck and administers the particular site. After this beginning the original founders of SHARE moved to different parts of the country, in part through an interest in expanding the network but also because of changes in their own lives. Distribution programs opened in Virginia, The South Bronx, Philadelphia, Colorado, Florida, and Milwaukee. However, the distribution model was always understood to be flexible and as SHARE sites opened across the country they partnered with different organizations and found distribution systems that worked for them. For example, SHARE in the Philadelphia metropolitan area asks for two hours of "good deed" time in order to receive food from the program, and the SHARE affiliate still active in San Diego does not require participation because of their concern that some people have physical or lifestyle restrictions, making such a time commitment impossible. Some SHARE sites partnered with the local food bank, CAD organizations, or chose to remain as stand-alone organizations.

SHARE is very much a no-nonsense organization that focuses on food access over specialty foods, price as opposed to packaging, and meeting people's needs as opposed to dictating the type of food that people should eat. The SHARE organization in Philadelphia uses the acronym ABC to describe their food: affordable, basic, and consistent. The SHARE sites in Duluth opened in 1985 and grew to serve around 30,000 people each month. Thus, like the Mondragon cooperative in Spain, they successfully solved the problem of "scaling up" community economy projects through the development of small and assertive leadership teams that were able to maintain commitment to the project over time (Gibson-Graham 2005). In addition to organizing monthly food distributions, they also started a mobile market that serves residents of food deserts in Milwaukee, and discussed the possibility of opening a retail store in a Milwaukee food desert. Chatterton and Pickerill (2010), in their work on "autonomous geographies" among activist groups, chart a similar discounting of ideological rigidity in favor of a politics of pragmatism in meeting the needs of everyday problems through an ethic of working together to solve communal problems. They observe that "political identities were constituted through the everyday practices of doing activism in particular projects and campaigns, rather than political identities pre-existing fully formed" (479). In this sense, the shape of SHARE and its ongoing changing reflect a politics of action over ideology.

A problem with SHARE's embrace of "everyday practices" is that the project of identifying the niche that SHARE would play in the food system was a terminal project as opposed to an ongoing relationship to the food system. To this end, many of the volunteers I met were similar to CHUM and RFP volunteers in that they were motived by action and "giving back to the community" as opposed to an ongoing project of community empowerment. In 2013, the economics of distributing food through a volunteer-led network changed and SHARE Milwaukee folded, leaving the local SHARE organization to decide how to stay in business. The larger Milwaukee-based SHARE stopped their distributions based on declining orders, which were understood to be a result of the recent increase in the number of discount grocers (such as ALDI, Walmart, and Save-A-Lot) and a change in buying habits of their customers, who became less interested in having their food choices limited to the 50 items available each month at SHARE. As one organizer noted, "SHARE filled the gap until the retail food industry began to consistently supply low-cost access to food; when we started it was either the food pantry or regular grocery store." Rather than change to a model in which they were dependent on grants, SHARE Wisconsin chose to shutdown.

The local SHARE community found a local distributor and decided to keep the name SHARE because of what it denoted and went about setting up a new marketing and distribution process. However, at this point the organization is on hold and they are working to see if they can continue to distribute food based on this model. As Dana described the problem:

> That word "share" has a connotation of need. I've spent the better part of eight, nine years explaining to people it's an everybody place. It's not just for people who need help. There are plenty of people who are regular folk who could do something with the deal... If people say this isn't convenient for us, we'll just go away. Then maybe we reinvent ourselves and we find a way to organize transportation for people to get the job done. We don't have to be just a food distribution thing through this warehouse. If there are people who are not eating well in this community because of those issues of insufficient resources financially or in transportation, maybe because we get how people move about the planet, we can then develop a transportation network.

SHARE formed originally to help provide low-cost groceries for victims of the 1980s depression in San Diego. As the grocery market diversified and low-cost foods proliferated, the unique niche that SHARE occupied left and the network is looking for a new way to reemerge in response to current conditions.

SHARE's core ideology is summed up in its name: Self-help and Resource Exchange. Their organizing principle comes straight out of the empowerment literature as theorized by Friedman (1992). Because their volunteers are mostly women, they reach out to this group in particular. Further, because women are most at risk for food insecurity, their focus is appropriate. The Hunger Project, for example, focuses on women in order to address food insecurity under the theory that this group is at highest risk for experiencing food insecurity. As noted in Chapter 3, they reach out to the highest earning group within the sample. Further, this was a group most able to provision for themselves and had the lowest rate of food insecurity.

Seeds of Success and the Struggle to Alleviate Poverty

CAD operates based on a framework of comprehensive engagement with individuals in order to transition them out of poverty. CAD adopts the most controversial stance towards workfarism noting that they do not facilitate poverty, but instead focus their work on moving people out of poverty. To this end, their 2013 annual report notes that in that year they served 382 people through their coaching program, and as a result 98% increased their net worth and 85% obtained a job. SoS engages with the community economy project through working one-on-one with individuals in order to empower them to participate fully within their communities. To this end, they decenter the neoliberal narrative that only strictly capitalist businesses can support communities and instead reframe the state as a strategic tool that can be utilized to improve individual livelihoods (Gibson-Graham 2005). Their long-term relationship with participants allowed for participants to leave SoS changed. In order to engage in this process, they draw on a wide variety of influences and work systematically to create a unique and holistic experience geared towards food security and transition out of poverty. This means that SoS has a very hands-on and pedagogical component designed to alter the behaviors and lifestyles of program participants. This is done through integrating the crewmembers with the other programs of Community Action, and helping them to understand things that they could be doing better in their life. Of the community-based food programs in this study, SoS is the most clear about articulating and naming the institutions that work against people experiencing food insecurity. However, their focus is on how people can thrive within this system through fully informed involvement with society, not on how to transform or reimagine neoliberalism. The project of Community Action Duluth is to see people living a different life after they complete the program and to provide all of the tools in order to make this possible.

SoS deals directly with a variety of health- and food access related problems: they sell their produce in a farmers market located strategically in a food desert, their urban agriculture project is part of a burgeoning local foods movement, they employ a SNAP outreach coordinator, and most of the participants in SoS changed their eating habits through involvement in the program. However, for SoS job training is always their first goal and these other interventions are secondary by-products of the overall focus. Paul, a program director, outlines the stress between these different goals in this way:

> Some of our funding at this point comes from [an organization] that's sort of a leader among funders [that is] attempting to suggest to people that your work should be centered around policy systems and environmental change and so, in a certain sense, there's continuity with the Catholic worker ideal doing the work of hospitality, but then also the work of organizing and activism on bigger national issues that you're not really addressing in your day-to-day hospitality. For us, we do some policy things related to urban agriculture and trying to push the envelope. The majority of our time is on the day-to-day work of providing a transitional employment experience and doing food access as work. I'd say that the greatest continuity there is in that duality of doing some activism and some direct service, or we'd probably call it environmental change.

Paul and SoS volunteers often attend local discussions about food access and health, but they have made the decision to fund their operation primarily through job training and urban redevelopment funding as opposed to developing an independent urban agriculture program. For example, their most efficient production location is on the outskirts of town, and they have chosen not to consolidate operations there to improve efficiently because it would take their operations out of the inner city.

The close supervision of SoS crewmembers is deeply reminiscent of neoliberal policing as explored by Rose (1999) within which the individual guidance and adjusting of mores is a technique of governance. Challenging this interpretation, Hébert and Mincyte (2014) argue that while "self-reliance" is championed by neoliberal scholars, it is also consistent with the survival strategies of people on the edges of capitalist empires struggling to provide themselves in spaces where cooperative and feudal economies support capitalist spaces. To this end, self-betterment for the purpose of better taking care of family is fundamentally different from the strategies pursued by the policing projects of the state. They argue, borrowing from the actually existing neoliberalism literature, that alternative economies are not sacrosanct or pure, but instead fluid and do not reject state and capitalism, but instead evolve in contestation with these other economies (Chatterton 2005; Chatterton and Pickerill 2010). Thus, working with individual crewmembers to make them more economically secure is a form of psychological and household empowerment in that it provided food-insecure individuals with the tools to improve the stability of the households (Friedmann 1992). As such, the hands-on training provided by Community Action Duluth is centered on the projects of *household* and *psychological* empowerment that gives individuals the skills to better interact with the political and economic causes of marginalization.

An example of this type of cooperative work is in relationships formed among SoS crewmembers as they cooperated on the labor of running the agriculture plots. They were not neoliberal automatons embodying self-reliance, but instead were proving childcare and cooking communal meals with one another (Hébert and Mincyte 2014). In this context of mixed economies, bringing state benefits in through SNAP benefits was not a contradiction but instead an extension of the fluid nature of actually existing autonomy. Thus, it was empowerment *through the process of collective work*. In fact, spending time at SoS you were constantly confronted with the ways in which their program imagined individuals behaving in different (improved) ways in order for them to live better lives. As Liz Bondi (2005) notes:

> Neoliberal subjectivity does not inevitably generate subjects oriented solely to the narcissistic gratification of individual desires via market opportunities. Indeed aspects of neoliberal subjectivity hold attractions for political activists because activism depends, at least to some extent, on belief in the existence of forms of subjectivity that enable people to make choices about their lives.
>
> (499)

Hence, in empowering clients, SoS can cynically be read as reinforcing neoliberal tendencies.

Crewmembers came to SoS at vulnerable times in their lives and were the most likely to experience some form of transformation during their time involved in the project. Samuel's story is typical. A 20-year-old African-American crewmember was forced to confront his shyness and nervousness at public interaction through his work facilitating the farmers market:

> [SoS has] helped me a lot. Not just the income, but also the knowledge. If I wouldn't have done the farmers market, I'd probably still be suffering from anxiety really bad. So like, since I started doing the farmers market, after the second time I haven't really had to take my anxiety medication. And it's like then we did the big one down at ... what is it? Canal Park and there was like 50 vendors, like 300 people, there's people everywhere. And it's like before I'd pretty much go crazy, like I'd go running, because large crowds and a lot of people coming up to talk to me and it's like, I did very well. Broke my anxiety barrier. Without Seeds I probably wouldn't have been put in a situation where I would have something like food to help me kind of talk. Before like, if I'd be in a crowd of people I don't even know what to say, but now I'm in a crowd of people, but I got something I know about that I helped grow and so it's easier to talk.

In addition to creating community, SoS put participants in new positions, helped them to confront the various barriers in their life and helped to transition them to new places where they were better off.

The mission statement for SoS, as Paul states, is three-pronged: "transitional employment to improve employment outcomes for long-term participants, improving healthy food access for long-time consumers, and for revitalizing neighborhoods through creation of green space." The comprehensive focus of SoS work is similar to the Te Runanga o Te Rarawa (TRoTR) project in New Zealand. Here, thanks to the hollowing out of the New Zealand state, the progressive NGO TRoTR that serves the Maori community has actually stepped into the physical space of the now empty social services building and become the exclusive provider of social welfare services for the region; they established their own "progressive" version of care under the larger state ideology of neoliberalism (Lewis et al. 2009). In this sense, their activist and empowerment-based project was established with state buy-in, indicating Keynesian and more progressive spatial and regulatory regimes can emerge at the local level within this era of downsizing (DeVerteuil et al. 2002). Lewis writes that this was not an event but a project of "persistent work to win privatized social welfare delivery and planning contracts and to integrate these into a coherent, strategic, locally driven development programme based on Maori principles and socio-spatial imaginaries" (168). The depth and size of TRoTR is reminiscent of the Keynesian state and the prison industrial complex (Gilmore 2007) but organized in a more progressive and empowering way. They liaise between citizens and the central government and offer health care and other social services. TroTR theorizes that gaining government contracts gives them expertise, and expertise can be used to integrate disparate projects into an integrated self-improvement program undertaken with state support and tailored to the needs of the Maori community. In light of Chapter 3's discussion of community, at TRoTR place-based identities and a coherent vision were utilized to bring the disparate members of the Maori community together through a strategic process of using the voluntary sector to effect state actions. SoS has also developed as a result of slow and intentional planning that aligned their interests in job training, community development, and individual transformation. SoS grew out of a discussion in 2009 between the Zeppa Foundation—a local foundation involved in alternative food systems and arts promotion—and CAD around a way to work together to create Green Jobs in Duluth. A grant was written to the USDA to fund a Community Food Program Research Grant that would involve studying the food needs of low-income consumers with the goal of producing a program that would create a local food program to address identified problems. The eventual results of this grant was a steering committee including representatives of different aspects of the local food community and a report including data from focus groups with low-income consumers and local producers. However, before the grant came in they decided to start the program using Duluth Youth Employment Service (DYES) workers, private foundation money, and economic stimulus money. This original version of SoS started in 2010 using youth employed for the summer through DYES to farm local vacant lots with 1/3 of the produce to be consumed by the crew, and 2/3 distributed to local wholesalers and retailers.

In 2012, as the amount of produce increased, they found that giving 1/3 of their production to the crew was not feasible, so the program continued wholesaling but looked for different models to use their locally grown produce to meet food access needs. As part of this process they increased donations to the food shelf and a local drop-in center, but out of an interest in developing a more market-based program they created Neighborhood Produce. This program was a subsidized CSA providing an assortment of produce grown by SoS to low-income seniors in Lincoln Park. This iteration of the program addressed the food access portion of their mission statement but only to a certain demographic. At this point, while employees were still coming to SoS through DYES, they also started using transitional employment grant funding to provide workers without a strong work background some of the soft skills needed to secure full-time employment.

Currently, SoS is the anchor tenant at a farmers market in the Lincoln Park neighborhood of Duluth. The market is open to all comers, and offers a 100% match for those using SNAP benefits and has a variety of other local producers selling their goods. The advantage of this new model is that it allows them to meet their food access goals while also reaching a wider demographic. As Paul explained:

> That's great for us because promoting something that's income limited is obviously problematic, because it's like, "Come, everybody, if you have income below a certain level," so that's a problem and this eliminates that and also eliminates the need for us to do any form of requiring people to state their income. It's just, if you have an EBT card, you're qualified to get this match.

This farmers market is open one afternoon a week and thus far has been successful at both moving produce and attracting a small number of other tenants.

The neighborhood revitalization portion of SoS's goal is met through turning vacant lots into active urban agriculture plots. In the 2012 growing season they farmed ten plots of land, but because the lots were small it required a lot of movement between the different sites. SoS also has space at a larger growing operation at the outskirts of town where they could conceivably centralize their growing operations, at the expense of neighborhood revitalization. Paul described this tension:

> I'd say over the last two seasons, sort of the general trend of things has been to try to centralize more. Apart from transitional employment and food access, we have a third leg of the stool, which is neighborhood revitalization through the creation of green space. So, we pulled back from some smaller garden sites that didn't really provide that function. Obviously, there's a lot of inefficiencies built into working in ten different places and logistical complications, but I can very much justify it if it's providing useful green space in otherwise messy and neglected lots, but I can't really justify it if it's

just in some church's backyard that will allow us to garden, if we were running it.

Driving through neighborhoods in Duluth, the SoS banners in front of vacant lands that have been turned into plots signify the revitalization work SoS is doing. Lessening these smaller plots in favor of consolidation would also destroy the conversations that take place between crewmembers and community members over what SoS is doing in urban neighborhoods.

SoS's mission of transforming people's lives means that they have a multi-pronged set of interventions into food insecurity that sets them apart from the other community-based food programs in this study. This commitment to comprehensive intervention means that they are comfortable talking with low-income people about new behaviors and actions that will help them succeed, and also comfortable reimaging the way CAD is run in order to ensure it meets the goals it sets out for itself. Rebecca described the relationship between food and health arguing that a real intervention into health and diet would mean the development of a health and fitness center inside a social service agency:

> I think processed food is the ruination of America. I really do. I don't know where to start. We had the extension service do a little cooking class. The person wasn't very zippy. I don't know, it just didn't click, but that kind of thing done in a different way is interesting to me. I'm really interested in health and I've put out a few feelers, if we're going to say we're a holistic agency providing holistic services, health and wellness I think have to be part of it in there and very little now is done here about that.

This type of change would be directly in line with ways in which CAD has evolved over time.

The Alternative Economy of Food-insecure Households

Participants at CHUM, RFP, SoS, and SHARE have complex provisioning systems ranging from relying on friends and family for food, gardening, fishing and self-provisioning as well as utilizing soup kitchens and SNAP benefits. In addition to using the strategies explored in Chapter 3 to portion and control their food, they also create a food system for themselves through strategic and assertive utilization of available resources. Framing this food system as one created by savvy and entrepreneurial individuals looking to provide for themselves and their families pushes back against the stigma associated with food insecurity and the alternative economy. As one CHUM volunteer noted, many participants at the food shelf are there because they are taking care of children, cousins, and grandparents, and they are involved in just the same kind of "care work" that volunteers are involved in. In this sense, the decision not to be an autonomous being responsible only for their own care is what has driven their use of the food shelf and engagement in the voluntary sector.

As Harris (2009) notes, many alternative food projects have neoliberal aspects but can also be read for difference to discern other aspects of their identity. Similarly, Holloway et al. (2007) note that when reading for alternatives to capitalism in alternative food projects scholars should embrace a heuristic methodology that recognizes that "there is no such thing as a singular alternative food economy, there are important discourses surrounding being different and doing this differently" (15). CHUM, RFP, SHARE and SoS each engage in a unique project of alterity, and their clients have all constructed a lifestyle for themselves that is in contact with neoliberal state institutions, but which also draws on their own families and unique way of being in the world. In this section, I draw on the voices of participants in these programs to explore how they engage in a process of reframing poverty and food insecurity and experience agency within the multiple economies that they touch on in their daily life. The importance of these provisioning strategies echoes research done in other food-insecure communities. Whiting and Ward (2010), for example, explore how residents of the Northern Cheyenne Native American reservation use a combination of cash, SNAP benefits, federal entitlement programs such as Disability and Social Security, community food resources such as hunting and gardening, and reliance on friends and family to provision themselves. They note that households that rely on community food resources, and family and friends for their sustenance experience less stress than those who rely on cash or government food programs, despite their higher levels of food insecurity. Similarly, Ford et al. (2013) note that food insecurity is correlated with not having a hunter in the family to supplement other food supplies, and meat from hunting is provided to food shelf users as part of their bundle of food. In this sense, reliance on unpredictable food networks derived from their community networks actually was less stressful than engagement with federal and tribal bureaucracy.

Families Shape the Contours of Food Access and Food Security

Families form a key support system for households experiencing food insecurity in Duluth and as a result food and other goods are traded within families in order to support the livelihood of multiple households. In the same way that volunteers provide unremunerated labor within the voluntary sector, people experiencing food insecurity perform care work for members of immediate and extended family. As Sarah, a 49-year-old Native American CHUM participant noted about her recent trip to the food shelf, "We have a lot of cousins who hardly have no food and now we can help them out too." Similarly, the mantra at RFP is if there is extra food it should be given to neighbors and family in need, and Marisa, a 72-year-old white RFP participant engaged in just this practice:

> INTERVIEWER: If you ever have too much food do you give it away or do you throw it out?
>
> MARISA: I do, I have. My two adult children I always run it by them. If I have extra or I won't be able to fit that in my freezer, "Could you take it?" We work it out like that, yes.

Similarly, food insecurity is confronted amid a web of other problems that low-income people encounter (Spencer 2004). Therefore, it is not only food that is traded between families, but also forms of support:

> INTERVIEWER: Can you give me an example of the last time you sought help from a family member?
>
> NANCY: Yes, the other day. I don't have any money again until tomorrow on payday. I asked my dad if I could borrow some gas money until Friday. He's like, "Why don't you just come work with me and you can have money?" I'm like, "Okay, deal."
>
> INTERVIEWER: Trade a little labor for some gas money?

The tally of economic activity within the rubric of the community economy project explicitly valorizes this labor; as Gibson-Graham note (2005) "the intricate *inter*dependencies of household, community, and market-based economic dynamics within household economies, the voluntary sector, or neighborhood economies is rendered virtually unthinkable by the hegemony of capitalocentrism" (58).

Reframing food-insecure individuals as caretakers within their own families is both a recursive project in terms of how the "whole economy" is understood, and also a question of recognizing the difficulty of performing the labor of social reproduction. Families, like all communities, are heterodox and contain individuals with different feelings and attitudes towards food, for example, based on generation and social class (Backett-Milburn et al. 2010). Thus, positing the household as a community devoid of difference overlooks the fact that for those in need these differences must be bridged and confronted. To this end, the everyday actions of social reproduction within families become the sites where food support programs are discussed and strategized around. Jennifer, a 48-year-old African-American CHUM participant, for example, notes that her daughter is too prideful to use CHUM even though she herself has relied on CHUM to provide for her family:

> My daughter. I be trying to tell her, "Girl your pride going to get you caught." My daughter is too proud to even step foot in a place like [the CHUM food shelf]. You was basically raised in a place like this ... What happened? Now she got her car, she going to school. Which I'm proud of her, don't get me wrong. She got two loving kids. She got her own place, she's doing good. Now she too good to come from the food shelf.

In this instance, family is the contradictory site where stigma and food access come together.

Musick and Wilson (2008) note that a key distinction between interfamily care work and traditional volunteer labor is that while it is expected that care work be performed, volunteer work is often pursued based on personal choice. Thus, fulfilling familial obligations does not allow for the same sense of autonomy and altruism that volunteering outside the home gives, but it still marks an intentional and time-consuming engagement with alternative economies. Judy, an RFP participant, explains that she cannot volunteer for RFP because she has so much stress helping her family out:

> Yeah, I'm just too much stressed right now to be ... I also help out my, I help out the family and try and keep the house running as smooth as possible, and I also help out my friend Susan with her husband. I go and stay there and help out because she doesn't get much sleep.

Similarly, James—an African-American CHUM client—connects his history of volunteer work at different Duluth-based charities to the concept of suffering, noting that his hardships inspire him to find the ability to give back and help others. As he noted:

> Okay. A lot of times they get food from [Loaves and Fishes], you know, I used to volunteer there, or go down there and cook food for them, something like that...
>
> INTERVIEWER: You have a lot of experience.
>
> ROBERT: Well you know I've got [a lot] in life. You know you still got to suffer everyday, you know what I mean?

It is important to locate family care as a vital component of how social reproduction operates in coordination with the web of social organizations that comprise the shadow-state and the voluntary sector (Wolch 1990; Fyfe and Milligan 2003). Supporting this project is as vital to the survival of food-insecure individuals as the operation of the shadow state.

While we tend to think about the term family as a motionless concept, this overlooks the fact that families are formed through migration and within family stories and memories are housed unique understanding of food and diet. Thus, in addition to serving as support systems to ward off food insecurity, they also provide guidelines for how to understand concepts like healthy and nutritious. For example, many African-Americans, with histories in the warmer climates of the South, talked about the beneficial aspects of food in a way formed by this past. Alan, a 47-year-old African-American CHUM participant constructs nutritious food in a historical and place-based context:

> If I just go back in time, it's your freshly grown and freshly prepared foods. Everything now comes through a processing plant and whatever, so you can only trust that people are making things healthy for us since we've got to go in the store and just pay cash for what we need off the shelf or out of the bins. It's just that trust factor. Nutritious, that's just what I think, fresh foods, like growing your own vegetables, having your own chickens, getting your own eggs fresh out of your own barn. Nutritious, to me, is just country living.

In the same way that food is shared within the family, this sense of trust also informs how products from the industrial food system should be understood.

Hettie, an elderly white RFP participant, highlights the interaction between stigma, food insecurity and social support. As a user of government benefits (SNAP benefits and Medicare) she spends a long time figuring out exactly which paperwork should go to each government office and has utilized a variety of food access programs in the voluntary sector:

> INTERVIEWER: Can you give me an example of the last time you asked help from a family member?
>
> HETTIE: Oh my brother, he kind of offered, so I didn't feel like I was [help], he killed a couple of cows and so he said "We have got a ton of meat, it's there for a long time, do you have any room?" So he didn't make me feel like "Can I please have some meat?"

Self-provisioning and Food Insecurity

As explored in Chapter 3, many people experiencing food insecurity saw gardening as an enjoyable activity and supported the goal of making this food source a more significant part of their provisioning strategy. As Table 4.1 summarizes, participants in each of the community-based food programs had complex alternative provisioning strategies and consumed food from a wide variety of sources in addition to the supermarket and the community-based food programs within which they participated. To this end, Ford et al. (2013) note the irony of providing low-cost commodity foods to food-insecure members of First Nations communities in the Arctic Circle when they would prefer gas to help them hunt or changes to hunting regulations in the Arctic Circle; the net effect is both poor nutrition and dependency (see also Harder et al. 2012).

While Duluth lies far from spaces traditionally associated with the local food movement, hunting and fishing are a strong component of life in Northern Minnesota and debates about the relationship between Native American fishing rights and the largely white sport-fishing community are a constant part of the public sphere (Silvern 2002). It is not uncommon to meet individuals with chest freezers full of venison and walleye, and for hunting to be understood as a way of lessening the overall food bill. In Minnesota, about 13% of the population hunt each year, and another 28% fish (US Fish and Wildlife Service/US Census Bureau 2007). While food provisioning in general tends to be dominated by women,

Table 4.1 Alternative Provisioning Strategies used by Participants in CHUM, Ruby's Food Pantry, and SHARE.

	CHUM	*RFP*	*SHARE*
< 5% of food from gardening	14%	26%	12%
< 5% of food meat from hunting	16%	19%	10%
< 5% of food from fish	12%	14%	4%

hunting and fishing are overwhelmingly male activities (91% and 75% respectively) and associated with nature conservation, spending time outdoors, bonding and camaraderie, as well as provisioning. As Eaton (2008) describes "hunting invokes an altered sense of consciousness, one of supreme alertness to the animal and the environment. It gets us out of ourselves, beyond our ego and as a consequence the hunt is fundamentally a religious experience, one that reconnects us to the source" (44). In the same way that community gardening is promoted for its ability to bring communities together through a shared outdoor experience (Ghose and Pettygrove 2014; Gibson-Graham et al. 2013), hunting strengthens the local food supply while also promoting recreational activity. A key distinction, however, is that while urban gardening demands a shared ownership of space and often involvement of civic-society organizations to organize and manage its development, hunting is usually self-organized and takes place on state-owned land and thus is less commonly associated with the development of an official organizing body (Duda et al. 2010).

Because of the low cost of participation for fishing and its recreational pull, fishing is an activity that people can be involved in with a minimal investment in time and income. Allen, for example, is a 47-year-old African-American CHUM participant who views fishing as an economical and enjoyable way to provide fish for his family:

> INTERVIEWER: Do you fish, or do you buy fish from the grocery store?
>
> ALLEN: Perch, most of your boneless fish. No one likes to struggle with bones, especially when you're feeding your family and you've got kids, excuse me. Perch was the best, perch, whiting. I got up here in the Northland, I learnt about tilapia, good fish. I never knew nothing about it. Then you get into the expensive fishes like the orange roughy, the walleye, good fish. I love fish, and I love fishing for fish, fresh fish.
>
> INTERVIEWER: Do you fish a lot?
>
> ALLEN: Not a lot, but when that time comes and I can get out there. I do enjoy fishing.

While gardening demands a constant commitment to upkeep and harvesting, fishing can be done on the timetable of the participant and combines time outside with nourishment. In a similar vein, James, an African-American CHUM participant, discussed fishing with a group of friends and noted that Minnesota, like other states offers reduced fees for fishing licenses for individuals with disabilities or receiving state benefits:

> You know, it was funny because I went fishing Sunday, you know, but I didn't fish because I had to go down there to go get this paper from the social security office so I could take it to the DNR so I can be able to get my license for free, and I was planning on doing that today. Good thing. When we went fishing, I seen this turtle. It was a big turtle, it was like this big [hold his arms

out wide]. Funny thing is the turtle walked up the steps, and I've never seen a turtle able to do that before. Then he came down stairs and it was a big old snapping turtle, you know this turtle could feed about 20 people.

INTERVIEWER: Oh my gosh. Could you eat it?

JAMES: Yeah, I could if I had to, but I had turtle soup before and it was in a can. Around here for somebody to fix turtle soup, not too many people experienced in that...

While there is a low bar of entry for fishing, to engage in hunting demands a firearms license, an in-depth understanding of the process of hunting, and access to state or private land upon which to hunt. Duda et al. (2010) note that "it takes a hunter to make a hunter" (67) highlighting the intergenerational transfer of hunting knowledge, and also framing hunting knowledge as a form of social capital unique to food-insecure communities outside of large metropolitan areas. This intimate project of knowledge-transfer points to the embodied and corporeal project of teaching hunting: only someone who engages firsthand in this process can become a hunter. Given the prevalence of reliance on natural resources as part of provisioning strategies, food-insecure individuals in Northern Minnesota present a relationship with the environment based on what Guha (2006) describes as "livelihood and survival" (63) wherein strategies of social reproduction and environmentalism are interwoven and care for the environment is linked to household survival. Hence, environmentalism is linked to appreciation for the environment that is a part of hunting, an appreciation for the ability of natural resources to provide for households and develop more sustainable forms of food access. As Robbins et al. (2008) write on the growth of Non-timber Forest Product (NTFP), "gatherers" in state-owned lands, "people are surreptitiously and personally involved in something very different from capitalist nature. For many, nature 'out there' seems connected to daily practice 'in here' in a relatively unproblematic way" (273). Understood through the lens of the community economy, natural resources are a commons from which livelihoods can be sustainably and ethically foraged, harvested, or hunted and food-insecure individuals can expand their pathways to food security while of course maintaining their connections to other provisioning strategies.

Reading Food Access as a Community Economy Project

CHUM, RFP, SHARE, and SoS use their position in the voluntary sector to meet the immediate needs of individuals experiencing food insecurity, but are also involved in a larger project within which they take steps to push back against neoliberal subjectivities and create an alterative space, or "spaces of hope" (Harvey 2000) where alternatives to capitalocentric values can survive. Levkoe (2011), in his critique of alternative food initiatives, argues that essential to mobilizing against neoliberal subjectivities is the development of "collective subjectivities," which support the development of solutions to problems in the present food system that "embed food within meaningful cultural and community relations while improving production of and access to good, healthy food for all" (692). Essential to this development, though, is an inclusion of "meaningful cultural and community relations" within the programs themselves, which is missing from many traditional anti-hunger programs. In a neoliberal era when both the right and the left recognize and glorify the position of civil society, actions can be interpreted in a variety of different ways (Rosol 2012; Mercer 1992) and collective subjectivities can only be fostered through concrete efforts to define and promote these ideals. Therefore, it is essential for alternative economy programs to be pro-active advocates for their ideas and not allow themselves to be constructed and understood (even by their users) as exemplars of ideologies that they oppose.

Each of the programs evolved over the course of their existence so much that it would be more accurate to define the programs based on their general form of operation and mission as opposed to their specific actions. Methodologically, this means that these are not case studies of static institutions, but instead studies of how the changing needs of those in need—and the difficulties of providing for them—force institutions to change and evolve in their operation over time. The flexibility within community-based food programs is similar to the zeitgeist of flexibility in late capitalism (Harvey 1991), but it is also distinct because changes were made not only in response to economic and governmental considerations, but as a result of ethical considerations as well as client need. For example, SHARE's goal of helping people had to change in order to compete with proliferation of new places to purchase food and new types of food that their volunteer-led distribution network could not compete with, and CHUM evolved from giving participants a box of food to allowing participants to "shop" for food in their small space.

Participants suggested a variety of different changes they would like to see brought into CHUM, RFP, and SHARE in order to better meet their needs (see Table 4.2). Gibson-Graham et al. (2013) discuss the empowering idea of a "tak[ing] back work" and "tak[ing] back the market" in order to make these paradigms serve the needs of the community. For programs that rely on donated food, evolution can be difficult. One RFP user commented on how the program could improve with "just a better [food] selection, but I know that they can't really control that. Because all that's donated right?" This reliance on donated materials shifts the burden of responsibility from the program organizers to the machinations of the neoliberal economy and disempowers this potentially empowering space. Here, I suggest that creating processes wherein program participants can expand the goals and objectives of existing voluntary sector programs is an example of "taking over the shadow state" wherein communities examine their strengths and needs and use these spaces to better serve their desires.

Using community food programs to improve food access as well as teach participants about food preparation, storage, and production further develops the problematic correlation between food and community. As Nancy, a 34-year-old white RFP participant noted about the idea of adding cooking classes to RFP: "I [would] love trying to have things like that. I don't have a lot of money for that, but if it was easy, that would be so fun, you know?" Similarly, Chester, a

Table 4.2 Other Services that Participants in CHUM, Ruby's Food Pantry, and SHARE Would Like to See Offered.

	CHUM	RFP	SHARE
Cooking with fresh ingredients	14%	22%	25%
Learning how to grow my own food	14%	24%	29%
Learning more about canning and preserving food	14%	34%	32%
Meeting others who use this organization	28%	12%	7%
Organizing carpools to the site	28%	14%	7%
Learning more about food support programs	14%	14%	13%
Learning more about government jobs programs	18%	12%	6%
Learning more about job opportunities	16%	18%	11%
Meeting local politicians	14%	7%	6%
Receiving health information	22%	28%	29%
Learning more about community organizations	20%	14%	19%

32-year-old African-American single parent and RFP participant discussed cooking classes as a way of having fun and drawing people together while they are waiting to pick up their food:

> That'd be great for those people who'd like to do it like me. I'd love to go in there, see, "Oh, hey, instead of using this like this, you can also use it doing this." Get more ideas for different meals to make, different things to try.

A difficulty, however, is that teaching, like charity, often traditionally involves the passage of knowledge from an empowered knower to a disempowered learner, and happens in a way that the pleasure and excitement of the information being passed is lost. As Bell Hooks notes, teaching "is meant to serve as a catalyst that calls everyone to become more and more engaged, to become active participants in learning" (1994, 11). Fred, a 30-year-old white RFP participant noted that these types of interventions must be done carefully so as not to stigmatize or speak down to participants but instead engage them in a way that moves beyond food:

> Right, so there will be one suggestion, so that's education dealing with food, nutrition...If you were to extend it further, I mean education on the program itself, maybe creating some form of empowerment so some people could share about what it is that the program is doing, to be able to expand in that way, that would work as well. Let me think in terms of other forms of education. It would be nice to see, I think everybody feels that they are intelligent and that they know what to eat, and yet at the same time the reality is we as Americans still eat poorly, if that makes sense. It will be wonderful to see a non-pedantic way of being able to remind people to deal with nutrition, if that makes sense.

Creating a joyous and "non-pedantic" way of discussing cooking and nutrition in the context of food access can only be done through a conscious inclusion of program participants and volunteers as a community of food-system researchers, interested in improving food access together.

Others built on Fred's notion of learning from other program participants about how the program worked for them, and discussed the need for job training, volunteer opportunities, and chances to better understand government programs. These calls build on the well-documented overlapping set of problems faced by food-insecure households (Watts and Bohle 1993; Whiting and Ward 2010). Marisa, a 72-year-old RFP participant noted that she was on waiting lists for government services, and saw RFP as a place where she could learn more about how to access the support she needed:

> I'm having trouble obtaining care for myself where I live. I guess maybe resources about that. I'm on a waiting list for respite care. I'm on a waiting list for a counselor. I'm having trouble finding a new doctor. It's just seeming like things are getting very short-staffed everywhere I live. My son was on a waiting list too and he couldn't get in at all for his own counseling and this has to do with his autism.

Given the resources put into the development and maintenance of programs like CHUM, RFP, and SHARE, Sampson, a CHUM participant, noted that they could provide just "the basics" about government programs to ensure that people were not falling through the cracks:

> I would say for those who, let's say we have some newcomers coming here to look, but they're not familiar with the rules and regulations. Help them out with programs to get them on their feet like going to apply for General Assistance, help them apply for General Assistance or if there's job training give them some resources they could follow up on to go to the seminars or go to the various employers that are hiring. Not too much, that's just basic.

Offering this type of assistance—something that CHUM and SoS already do— would allow food access programs to assist their participants in achieving food security and start conversations that could lead to other forms of activism and program involvement. Importantly, they would also respond to the voices of program participants representing their views for how these spaces can work for them and fulfill their needs.

Learning from Alternative Spaces

The incredibly hard work performed by volunteers and staff members at food access organizations belies the myth of revanchist urbanism (Smith 1990) and the prevailing story of the city as a harsh and unwelcoming neoliberal space. CHUM, RFP, and SHARE tell the story of volunteers working collectively to ensure that those in need receive the help that they require; they also tell the story of the development of unique and vibrant spaces of care and hope (Harvey 2005). The diverse ways in which alternative economies developed within community-based food programs and the lives of the food-insecure individual suggest novel ways in which community food programs in particular and institutions in the shadow state generally can challenge capitalocentric thinking.

Each of the projects struggle with the notion of scale and how to link the work of meeting the immediate needs of those in need with the larger political project of empowering the hungry. CHUM's unique structure as a coalition of 40 faith-based institutions that connects improved food access with political advocacy for those with low or no income "jumps scales" through its embrace of progressive state-level policies. As an organization closely linked to the state, this activism challenges the inherent complicity of shadow-state institutions and suggests that this space can be reformed from the inside. Similarly, the thoughtful engagement with individuals at SoS pushes back against the neoliberal state's disciplinary framing of citizenship in favor of an understanding of personal responsibility as a necessary precursor to transformative action. Similarly, the RFP and SHARE leadership teams transformed individuals with little experience in organization or activism into experts of the particular task of program management.

The challenging of neoliberal subjectivities took place in different ways at each of the projects. SHARE consciously supported the development of "collective subjectivities" through the creation of buying clubs and pushed back against stigma through naming program participants "consumers" who used the market to support their lifestyle. SoS, in particular, has a comprehensive commitment to transforming participants' lives through one-on-one mentoring in order to not facilitate poverty. In this individual-oriented work, they recognize the connection between neoliberal subjectivities and the development of skills that promote the emancipatory potential of individuals.

The lack of formula for how to incorporate the alternative economy into the shadow state speaks for the need to think creatively and comprehensively about how to tie organizational mission to the expressed needs of the community. The coordinating committees at SHARE and RFP, for example, operate in a manner similar to "autonomous spaces" where individuals with little experience at activism could work together to develop new models of change. Likewise, religion played a strong role in the organizational culture of CHUM and RFP, but did not affect the lives of participants nearly as strongly. As discussed in the previous chapter, this contributed to the development of stark differences between volunteers and program participants.

Individual participants in the programs have complex and innovative food management and procurement strategies that utilize particular place-based knowledges. For example, in a city with a strong outdoor sports community, hunting and fishing formed an important part of many household diets, and these practices also challenged neoliberal subjectivities. Pairing these emancipatory possibilities with the longer list of programs that participants would like to see added to CHUM, RFP, and SHARE support the contention that taken collectively, these programs inhabit spaces that are much more elastic than simply agents of the hollowed-out state. This insight is especially important because of the larger neoliberal and workfarist state that surrounds and informs much of the work of CHUM, RFP, SHARE and SoS. The next chapter examines the larger neoliberal context within which these modest disruptions take place. Here, I note that while the organizational mission and personal motivations of volunteers may reframe hunger, these same actors are forced to ration food, set expectations on participant behavior, and collaborate with the neoliberal state. Thus, their discursive and material practices that challenge capitalocentric logics are stymied by the larger milieu in which they operate. Further, because people experiencing food insecurity must receive aid from multiple governmental and shadow-state sources, these overlapping sets of regulations mean that they are measured and ordered by a dispersed set of institutions that impede their ability to obtain a higher level of food security.

Notes

1   From *Catholic Radicalism* by Peter Maurin. "The Fallacy of Saving," 1949.
2   From *The Conquest of Bread* by Peter Kropotkin. Page 281.

5 Neoliberalism and the Porous Continuum of Care for the Food Insecure

> *This is the secret of wealth: find the starving and destitute, pay them half a crown and make them produce five shillings worth in the day, amass a fortune by these means and then increase it by some lucky speculation, made with the help of the State.*[1]
>
> *Right now, yes. She's only been gone for not even a week. I already miss her. Let's see, Thursday, Friday, Saturday, Sunday, Monday, Tuesday, it's been seven days she's been gone. I can't wait to see her. Actually, [...] she's up there eating three meals a day anyways.*
>
> Victor, CHUM Food Shelf participant,
> discussing his girlfriend in the county jail

The variety of problems faced by the people who arrive at the CHUM Food Shelf on a typical day underscore the frayed state of the existing social safety net and the difficulties inherent in having organizations in the voluntary sector provide emergency aid. A newly arrived family staying at a hotel requests food that can be prepared with minimal kitchen facilities; a young family without soap, toothbrushes, or pots and pans asks if any of these items are available; men staying at the homeless shelter across the street need food items that they can store and eat easily; a woman recently released from the hospital asks if we can help her find food that won't aggravate a chronic stomach ailment; and, like clockwork, those on state benefits request enough food to cover until the next payment arrives. Inevitably, clients are juggling the demands of family and work obligations and often utilize a variety of state aid programs to provide for their households. Volunteers and food shelf staff meet clients as they arrive and help them fill out the required paperwork asking about sources of income, use of state aid programs, family size, frequency of food shelf usage, and how they have arrived at the food shelf. The charity-minded people who donate their time to the food shelf for altruistic reasons must act in the capacity of bureaucrats as they collect data, ration scarce food stores, and address the withering complexities of individual circumstances. In this capacity they are inevitably put in the lordly position of helping to discern how much aid to give, where to refer people for further support, and what benefit programs may exist to refer them to.

This chapter addresses the relationship between workfarism, neoliberal welfare policy and shadow-state institutions. It makes two arguments. First, I contend that each of the food access organizations studied have distinct workfarist and neoliberal tendencies that, consistent with Nikolas Rose's (1999) conceptualization of neoliberal subjectivities, position participants as "deficient" and in need of appropriate training. For example, as explored in Chapter 3 their food provisioning strategies are not valorized. At SoS, they are explicitly trained to better participate in the workforce, and at RFP they are offered a "hand up, not a hand out" in order to reduce the chance of dependency on charity care. Second, the persistent workfarist tendencies of the organizations are particularly problematic because of the precarious position that food-insecure individuals are placed in as they interact with the web of welfare organizations that serve the poor. As Dorothy Day (1996) writes, "how terrible a thing it is when the state takes over the poor! 'State ownership of the indigent,' one of the bishops called it" (19). I argue that these agencies form a "porous continuum of care," which positions those in need of food in an ineffable position: they are dependent on state programs for health care, economic, support, food support, and other benefits. However, the logistics of receiving care are so onerous that participants do not have the ability to transition out of poverty.

Community-based food programs exist in a complex position vis-à-vis the state: they are at once collaborators with the state, vocal opponents of state policies, and attempting to build an alternative system of support apart from the workfarist state. As Berner et al. (2008) note in their analysis of participants at an Iowa food shelf, because of low wages and the porous nature of the welfare state those experiencing food insecurity exist in a liminal space between state welfare policies and employment; 39% of food shelf clients receive government benefits but must supplement these payments with the food shelf, while another 25% are employed but still in need of assistance. CHUM and SoS are mostly directly connected to the state: The Emergency Food Shelf network was created in direct response to neoliberal rollbacks of food support programs for the poor (Hawkes and Webster 2000) and Community Action agencies were created as part of the War on Poverty and receive a constantly diminishing amount of federal funding to engage in the project of social development (Orleck and Hazirjian 2011; Conlan 1998). Although RFP views itself as distinct from the state and describes itself as a benefit program for those who earn too much to receive state benefits, it receives food donations as a result of generous state support for commodity crop production and the tax write-off that corporations receive for charitable donations. In addition, each of the organizations have clear neoliberal and workfarist tendencies. For example, SoS has at its core the role of a job-training project; CHUM doles out limited amounts of provisions and coordinates with other area food shelves in order to forestall dependency; RFP considers itself a support program for those above poverty status who have been "missed by the state" in order to purposely distance themselves from other state welfare programs; and SHARE engages in a process that transforms food-insecure individuals into empowered economic citizens purchasing food through astute use of the market (see Rankin 2001 for the connections between these types of programs and governmentality).

Participants in community-based food programs also have a complicated relationship with the neoliberal state: although many depend on state and community-based programs for survival, they are also struggling to move away from these programs and improve their employment situation. However, these moves are complicated by chronic health conditions, difficult life situations, and the time-consuming nature of survival on meager incomes. The stigma associated with state programs and the difficulty of patching together different government and non-profit benefit programs complicate this process (Link and Phelan 2001). In sum, the struggle for survival takes precedence over effective mobilization to change their life conditions. While many individuals may embrace the metanarrative of neoliberalism, they also recognize the difficulties within their own life that prevent them from reaching this ideal of autonomy as they rely on their friends and family for support.

Further, the necessity for people experiencing very low food security to utilize multiple state and civil society support programs to survive interacts with the onerous protocols at each program to create a "porous continuum of care" within which the hungry are shunted from agency to agency in order obtain the elements needed to survive but treated in a manner that ostracizes them from society and impedes their ability escape poverty. The contention that the devolution of welfare policy to the state level has created a diffuse and interlacing network of actors that subjugates the poor is not new: Michelle Alexander (2012) argues that various aspects of the criminal justice system work together to create a restructured and renamed version of Jim Crow, and Loïc Wacquant makes a similar argument about the rise of workfare and prisonfare as a way of controlling "the precarious factions of the postindustrial working class" (2009b). Wacquant, building on the work of Piven and Cloward (1971) argues that "*the misery of American welfare and the grandeur of American prisonfare at century's turn are the two sides of the same political coin* (203, italics in the original)." That is, the same low-income, poorly educated, disproportionately people of color forced off welfare and into the shadow state are the same individuals subject to mass incarceration and the rise of the prison industry (see also Gilmore 2007). Lynne Haney (2004), perceptively argues that the devolution of poor care to multiple agencies and institutions does not necessarily translate into a coherent (or competent) system of care. Instead, she notes that the "the diversification of institutional practices has reached a point where subjects are governed through ambiguity and contradiction. So although the overall effects of this governance may be patterned, the actual processes of intervention are fraught with inconsistency" (349). In the case of people experiencing food insecurity, they confront multiple state and voluntary agencies with different philosophies of service provision and poor integration that together form a porous continuum of care and geography of isolation as people attempt to navigate these different systems.

Workfarist Regimes in Community-based Food Programs

The neoliberal state plays a powerful role in the creation of food insecurity: USDA policy creates overproduction and sets the rules for how SNAP benefits can be doled out. But these connections are rarely exposed in the concrete act of volunteering and fulfilling the biblical imperative to serve those in need, instead community empowerment is presented as a project of engaged community members instead of a "technology of governance" (Keil 2002). To this end , Wolch (1990) notes that economic inequality is a key part of the formation of the shadow state: "with the expansion of social problems connected to industrial capitalism, the [voluntary] sector assumed a key role in the implementation of antistatist political programs advocated by capital" (78). Rather than disappearing, for people in poverty the state is an intrusive part of their life providing services through a compartmentalized set of agencies that coordinate with one another and with non-state actors, such as community-based food programs, and ultimately weave together a porous and time-consuming continuum of care.

CHUM: Enumerating and Serving the Hungry

CHUM is part of the Emergency Food Shelf system and is on the frontline of meeting the needs of those experiencing food insecurity in Duluth. While a non-profit organization, it has many of the trappings of an official branch of government. In 2013, the food shelf served roughly 6,000 individuals, about 7% of the population of Duluth. In order to meet these needs they rely on fundraising—from the member churches of CHUM as well as in the wider community. The food shelf has set hours: one afternoon a week they stay open late to serve working families, and they are open five days a week to serve anyone who comes in their door.

Numbers and counting are key to the smooth operation of the food shelf. As Nikolas Rose (1999) writes about the relationship between numbers and democratic governance, "they redraw the boundaries between politics and objectivity by purporting to act as automatic technical mechanisms for making judgments, prioritizing problems, and allocating scarce resources" (198). They are a technique of governance that reduces complexity and creates the distinct categories of counter and counted. In this depersonalized process, "new conduits of power are brought into being between those who wish to exercise power and those over whom they wish to exercise it" (1991, 676). In the case of CHUM, numbers allocate food to those in need and separate volunteers from clients. Everyone who enters the food shelf in need of food is asked to fill out a form asking questions such as permanent address, number of people in the household, place of employment, amount of state benefits received, and monthly income. Each time a participant returns to the food shelf, this information is updated and either volunteers or the food shelf coordinator talks to the participant about what benefits they might be eligible to sign up for or take advantage of such as SNAP benefits or WIC. A note-card system is used to keep track of this data—a person is employed to enter this data into a spreadsheet, and a can of sharp pencils is on hand so clients can fill out the worksheet and volunteers fill in the data. Participants are welcome to use the food shelf once a month and are never denied if they are desperate, but the policy is to encourage participants to find a sustainable way to make ends meet without coming more than once a month. In addition, the food shelf coordinates with the other local food shelf to ensure that participants are not relying on food from multiple locations in order to avoid finding a more sustainable source of food.

The data-heavy space of the food shelf continues to the selection of food. The food shelf is laid out in the client choice model. Under this system soups, meats, dairy, eggs, and other types of food are grouped together and clients—depending on the size of their family—have the ability to choose how much of each food they wish to take. This becomes the organizing rubric for the relationship between clients and volunteers. As Lindsey, a food shelf client, describes it, "You get a lot of points for veggies and then it seems like a lot of points for fruits. A couple of apples is a lot of points. They give you points and that's like dollars or whatever."

The collection of data and numerical-based rationing of food at CHUM is a stigmatizing project that is necessary because of the high level of need for food assistance in Duluth and stringent USDA requirements. As a condition of distributing TEFAP foods, the USDA requires proof that the food is going to low-income people. Therefore, CHUM keeps a database of food shelf users to prove the eligibility of food recipients. In addition, CHUM collects in-depth demographic and economic information in order to screen food shelf recipients for state-benefit programs. Because CHUM serves so many clients each month and are reliant on donations to purchase food from the food bank, screening for state benefits and coordinating with other food shelves to avoid participants using multiple food shelves is a necessity imposed by USDA regulations and their limited food supply. In this sense, CHUM collects data and allots food because of their parastatal status. This removes the state from the direct process of limiting food access by off-loading this project to religiously motivated volunteers.

The reign of numeracy at the food shelf is part of the clinical and formal process of intake that structures the relationship between volunteers and participants and ensures that these two groups relate in a professional as opposed to a communal manner. There is an implicit and explicit understanding that using the food shelf is a source of stigma, therefore those in need should find another way to feed themselves and participants should be treated with respect during their visit. For example, Cindy noted that to protect privacy they tend to use numbers and treat participants as though they were "co-equals" as opposed to close friends:

> I try to make it a friendly, comfortable experience for people rather than, that we are more "co-equals." I mean, I know we can't be "co-equals" but that I am a sympathetic listener and I can give them the time to tell me what their issues are and I can affirm them. I can give them a referral if I think that's appropriate and interestingly—because people don't want to be there for the most part and with the economy the way it is—we have found more people

who have been donors to the food shelf and all of a sudden they are "the others" and they are humiliated, they are ashamed, and you just need the time to be kind to them and help them understand that it's okay and that it won't be forever, without giving them false hope.

Tracy, a 62-year-old white minister and food shelf volunteer talked about the distance between volunteers and participants noting that creating close relationships is difficult because people are using the food shelf during a time of great need:

> I guess partly it is the realization that there are boundaries to my help, I mean sometimes we literally joke, I mean to say, "What extra do you need?" and they'll say "Oh well, that lobster dinner for 100 dollars would be fine" and I'd be like "Oh, shoot we just gave that last bit away!" or "Oh man, wrong door!"

The message to participants from CHUM is that the food shelf should be their backup, but not permanent food supply, and that the food shelf would like to encourage them to find other ways to feed themselves in a sustainable way. By counting and enumerating participants, CHUM is able to meet the needs of funders, such as the Federal government, that require that overstock food is distributed to those at, or near, poverty. These tendencies include a focus on consumer choice over food access, local control as opposed to national oversight, and weigh-ins and mandatory BMI check-ins for school lunch recipients (Allen and Guthman 2006).

Follow-up services with CHUM participants are provided through other branches of CHUM, and the food shelf serves primarily as a frontline for those in need. The absence of the CHUM Food Shelf from playing a long-term role in people's lives is in contrast to the theology of care and support for the marginalized that informs much of the rest of their work. Similarly, by recording the low incomes of participants and parceling out enough food for the next week, the food shelf feels like a disciplinary institution, which documents and monitors the actions of the poor. Williams (2012) notes that while 100% of African-Americans were disciplined through the Jim Crow system in the south, a much smaller percentage caught up in the New Jim Crow of the criminal justice system. However, the widespread use of surveillance, policing, stigma, and racism work together to create a still effective patchwork of ideological oriented institutions to continue this race-based caste system.

The ambiguity of the point system means that volunteers have the power to determine exactly how to count items and can allocate more or less food depending on their mood, read of the client, or simply willingness to adhere to the rules as outlined. This process is similar to Lipsky's (1980) recognition that street-level volunteers structure how institutions relate to individuals. Linda describes how she used this discretion, but also how participants self-police and accept this allocation:

> They do, [ask for more food than allotted] I mean to a degree. "Oh, this is 3 points, I only have 2 points left. Can I please, can I please, please have it?"

and that kind of stuff I usually personally don't care about. But it's more other folks that are miscounting or being sneaky as opposed to being friendly, I guess. But the vast majority are just so appreciative and the point system that we have they follow it to a T, and they are very careful not to take more than they are allowed for their family size and things like that. So most of them are like that, very appreciative. In my crew, with working families, I can say that they are just hardworking folks.

Samuel, a young SoS crewmember, saw the food shelf as a place with a negative stigma associated with it, which was nevertheless "a necessary evil." He used the food shelf in order to transition to SNAP, which would provide a stable and secure source of food. As he explained:

Well, I figured since we ... we started getting SNAP, we really didn't have the need to go to the food shelf and I'd rather somebody else who really needs it goes than me, because we do have an adequate amount of food. There's never a point where there's nothing in the house to eat. And I'd rather leave the services to somebody that could actually use it more than me. With the SNAP it does make it a lot easier to be prepared for the whole month.

To Samuel, although both SNAP benefits and CHUM have stigma associated with them, the transition to SNAP benefits brought a sense of security absent at CHUM.

At CHUM, scamming was something that not only affected the relationship between program participants and volunteers, but also affected how program participants viewed one another. As Bee (2011) notes in her analysis of the policing of others in loan circles in Bolivia, "while microfinance seeks to create communities of female borrowers who provide each other with social and financial support, these communities become a source of tension as poor women are made responsible for each other's debts" (32). Because rules at the distribution sites were not always enforced uniformly, participants were unhappy when other participants got preferential treatment, or when they saw other participants taking more than they deserved. Because food at each of these programs is in limited quantity, it was in the participant's best interest to make sure other participants were not scamming because they saw themselves as hurt by this process. Unequal enforcement of rules also resulted in a perception of the programs as being unprofessional or not up to the task of ethically and fairly distributing the food that was entrusted to them. Scamming, therefore, was not only about food quantity but also the overall competence and professionalism of the program.

Katrin, a food shelf user, unfavorably compared the lax policing of the food at CHUM to the more stringent food shelf requirements at Chicago-area food shelves that she had used in the past. She argued that the food shelf could be run better if they were more on the lookout for people who are getting food to sell on the street, not to actually feed themselves and their family. She argued:

> The people that come in, they don't get the food to take home. They ask if you can find food that they can sell on the streets. They should just let people know if you ever get food from here to use it because there are people out here, like I'm from Chicago, they don't have this kind of place there. A lot of people are hungry and kids that are not eating. But they take food for their [own] selfish reasons. I would actually screen them a lot better.

Because food is a scarce commodity, scamming affects how much food is available for Katrin, and reflects poorly on how CHUM administers its limited supply of food.

Because food is rationed at CHUM and volunteers are the gatekeepers, there is a natural tendency towards scamming and finding ways of getting more food by using multiple food shelves, asking for food for non-existent children/dependents or simply looking to get more food for the available points. Cindy, a food shelf volunteer, described how she double-checked the story of a potential food shelf user before giving them food:

> I had a family come in and they told me that there was a husband and a wife and they said there were eight children and they gave me all the information and they said they were temporarily staying at the drop-in center [across the street], but that they needed food and that they were getting into an apartment the next day. So when I went and checked with [the food shelf director], she checked across the street and there was no family of ten staying there at the drop-in and what we found out was that the mother and father were here, but the children were all still in North Carolina. So we had to give them the choice to either take food for ten people and not come back for a month, or take food for two and return [sooner].

In ferreting out this potential scam, Cindy is placed in the position of checking and validating what she hears from potential food shelf users, and has to control the ability of others to gain access to this food. The policing of clients by volunteers and clients of other clients is a specific technology of governance (Keil 2002) that emerges in the form of a "pattern of intervention" (Macleavy 2007), which shapes how individuals relate to one another in the voluntary sector. This watching and guiding of others is not a unique occurrence but instead an essential aspect of the space.

One result of the rules is a feeling among program users that CHUM staff and volunteers are nice "in spite of the rules" or that there is set of policies that can be changed or elided with proper behavior or demeanor. As Victor, a 37-year-old Native American food shelf participant explained:

> Staff here, they're very friendly. They've always been very nice to me. Even when I haven't had the, I forgot my ID one time, they still let me go through. The next time I had my ID but not my social security card, they let you go through again. They only can let that happen a couple times. The next time, if

you don't have the proper stuff there, they won't let you go through and they are sorry about it. They know you do need the food and stuff. They have to follow the rules also. All the food places are like that. I think they're very nice.

More important than how much food was actually preserved for use by others by the enforcement of these rules, they pitted volunteers against participants and participants against participants as each group witnessed the enforcement of rules and thought of ways to circumvent the rules in place.

Ruby's Food Pantry: A Hand Up, Not a Hand Out

RFP sees itself as a program that meets the needs of those who are experiencing food insecurity through a process that promotes self-respect and the ability of individuals to provide for themselves without experiencing the stigma associated with other food access programs. Participants arrive and purchase a $15 share from a volunteer and are given a number to use as they walk through the line. If an individual does not have the funds to purchase a share, they are asked to talk to one of the coordinators and there is some leeway allowed to fund people without $15 to also participate. Given the necessity of paying $15 to receive a share of food and the "hand up not a hand out" mantra of the organization, RFP is similar to the concept of "pious neoliberalism" within which neoliberal subjectivities are created in tandem with calls for faith, volunteerism, and conservative Islamic ethics (Atia 2012). In this Christian context, pious neoliberalism is enacted not through a focus on rewards in the afterlife, but within the context of being pious through actions that promote entrepreneurship and accountability among the pious. Although many participants might benefit from a screening for state-benefit program as happens at CHUM, participants are not screened for state benefits, instead they proceed to a sanctuary to wait for their chance to go downstairs and collect their food. As Rose (1999) writes, "control is now to operate through the rational reconstruction of the will, and the habits of independence, life planning, self-improvement, autonomous life conduct, so that the individual can be re-inserted into family, work and consumption, and hence into the continuous circuits and flows of control society (270)." In this context, the freedom from benefits and the ability to provide for oneself through the generosity of a faith-based project is defined as better than reliance on state largesse. In general, the distribution site is much more cheerful than the food shelf, and there is a feeling of people arriving for a deeply discounted sale from which they will emerge with something valuable.

RFP's lack of income verification and wish to meet the needs of all who have arrived is important to organizers, and many members of the organizing committee saw equality based on shared $15 investment as an essential aspect of the program. Ellen, for example, talked about how by eliminating guidelines they provided more self-respect that CHUM did. As she expressed:

> Well, I just don't like the idea that in a country like ours people are going hungry and I am very, very frustrated that low-income people have to jump

through so many hoops to get stuff. I mean, I find that appalling and I think the food shelf is great, but you can only go there once or twice a month and blah, blah, blah. The other thing I like about Ruby's Pantry is that I truly believe, if people pay something they have more of a vested interest and they have retained some of their self-respect. They don't feel like they are getting charity. They feel like they are getting a deal, it's a whole different thought.

RFP is set up on a model of market equality where the purchase of one share allows a participant to receive the same amount of food as anyone else. In this sense, waiting for food and purchasing (albeit at a discounted rate) illustrates the entrepreneurial initiative among participants, and the linking of self-aware economic activity and moral uprightness. This served to isolate RFP participants from the idea of state responsibility for food security and codify civil society as the correct place for anti-hunger work. Similar to the "pious neoliberal" organizations Atia (2012) analyses in Cairo, volunteerism and entrepreneurial activity are the correct response to poverty as opposed to activism against the state. Consistent with the powerful idea of community at RFP, organizers assume that people will eat the food themselves, others will share it with neighbors, and still others will give the food away to the food shelf or to needy individuals that they know. Bernadette, an RFP participant, analyzed this connection be describing RFP as the American way of providing food support:

> I think there's more of a humbling feeling for the people who use the food shelves than there is a stigma attached [to using it] by other people [...] to the ones who are going to the food shelf. Where like with Ruby's Pantry nobody says anything because you're paying for your food. You have a sense of pride about it and I mean it's the American way. If you can get a good deal you go for it you know. [laughs].

RFP embraces the American way through the notion that food choices should belong with the individual.

Using the $15 share price as the only arbiter for how food is distributed is understood by many organizers as a way to distinguish themselves from other food programs in the area. The $15 operates a screening system to ensure participants have a buy-in to the program, and a way to ensure that they serve a different and slightly more affluent audience than the food shelf. As Nathan compared RFP to other programs:

> I think each one has its place. I feel that for some of those programs, as any programs, they are taken advantage of, but how do you police that or how do you determine who is most deserving? How does one make that determination because a lot of the other programs, Salvation Army's food distribution program for example, I think CHUM is that way too, you are allowed this much six times a year, so come every other month. So what about the other months? I guess that's why we are different at Ruby's. We do not have any

income limitations, we don't have any frequency limitations, we do not have any quantity limitations, meaning if you wanted to come in and get more than one share we just ask that your contribution be equal to the shares that you wanted to get. Where with the other programs you get two weeks or three weeks of groceries, I don't know what those guys do but it goes back to how does one make the determination that you are worthy but you are not? You are not because you make just a little bit too much money?

By using phrases like a "a hand up not a hand out" and asserting that they serve people who make "just a little bit too much money…" RFP has constructed an alternative place for itself as distinct from the state and distinct from means-tested social safety net programs that they feel breed dependency and inaction. Instead, straight charity is eschewed in favor of a market-based understanding of individual self-dignity.

The importance of autonomy, paying for food, the appropriateness of civil society responses to hunger, and remaining distinct from the state is present both among organizers and participants at RFP. In the same way that the threat of scamming and the idea that food is scarce and therefore the object of conflict between those in need, at RFP participants view themselves as different from food shelf users. As Aliina put this distinction:

> It does give you a little bit more sense of self-reliance and pride to able to [buy food]. Even though you're getting like $75 or $100 worth of food for the $15 you pay you feel better about the fact that you help contribute to filling your larder. That's why I think a lot of people go there is because it's a matter of pride.

SHARE: Collective Consumption for Self-improvement

At SHARE, organizers are interested in transforming individual food consumers into collective consumers who understand that through working together they can stretch their food dollar and better provide for themselves and their families. As Bondi (2005) notes, there is an inherent tension between empowerment as an ideal of neoliberalism and market-based freedom, and empowerment as a project that has the goal of uplifting "people whose capacity for effective agency has been eroded by inequalities and oppressions," (506) and is thus focused on exposing the processes of disenfranchisement and providing participants with the skills to circumvent these obstacles. As Bee (2011) notes, the ideal of personal empowerment created through lending circles of poor women becomes more similar to community prisons if women are unable to pay back their loans, and social ties within the community are used to force/coerce payment. Like RFP, SHARE consciously differentiates itself from charity programs that do not require payment or volunteer labor from participants and view the fact that participants are paying to participate in the program as a distinct benefit. Participants are not screened for the eligibility for other state benefits, and because there are no income requirements for participation, the goal—like at RFP—is to sign up more participants. In fact, the more people who participate in SHARE the better, as it raises their collective purchasing power. While RFP asserts that people who make just a little too much money to qualify for state benefits are hurt under current guidelines, SHARE is focused on helping participants engage meaningfully in the existing market economy. As analyzed in Chapters 2 and 3, at the level of the specific SHARE sites the program was not particularly successful at building community among volunteers or in articulating a clear alternative vision for how their alternative economy operated.

A key to the long-term survival of SHARE is educating people about the power of collective consumption and differentiating their distribution process from other local food programs. For example, as RFP became more popular in Duluth, local organizers expressed concern that people's understanding of value would be skewed as they saw people returning from RFP with mountains of food for only $15. This forced them to be clearer about articulating the benefits of SHARE such as the higher quality of food and the ability of consumers to choose what their food package would include. This project of educating consumers about their role in the market is at once a question of transforming individual consumers into collective consumers, and at once a question of ensuring those who use SHARE that they are not part of a charity food distribution program. This is difficult because the name SHARE sounds similar to the acronyms used to describe other subsidized food programs (like MFAP and NAPS), and the non-traditional nature of the program means that people are going out of their way to use a program and not a supermarket. Dana phrased the dilemma in this way:

> There's some people that call me and they'll say, "I'm not broke and I don't want to shop with SHARE because then I'll reduce the amount that's available for somebody else." I'll think, "Okay. Let's take five minutes and talk about that. It's not a program. It's not like we get a fixed number of these things and once it's gone, it's gone. It's a buying club. In fact, the more people who buy, the better it is because then the people who shop at the warehouse level that we shop from them, they can negotiate a better price because there's a bigger volume of food that's passing through that channel." [And they reply,] "I didn't know that."

When volunteers are boxing up the orders for participants they always dump the food out on the table in front of them and say "this is your food that you paid for, so I need to go through and make sure your whole order is here…" This ritual ensures that the participant understands that they are not participating in a charity, but because it is performed by a volunteer (and not a paid employee as would be the norm in a regular supermarket) it reifies the liminal space of SHARE.

While RFP values individual autonomy of the participants, at SHARE collective action is prized. The choice to use existing civic institutions to host SHARE sites is an expression of this idea that community organizations in low-income areas should be supported, and this effort consciously pushes back against the ideology of individualism. Volunteers did recognize the collective power of the institution, as Sharon, a 68 year old working in the healthcare industry expressed it:

> As a worker, I just see us coming together with the community to help out in whatever way we can, as far as the food goes. I would like to be able to offer other things too, like toilet paper, paper towels, napkins, whatever that they can't buy on food stamps.

SHARE's promotion of the power of collective agency within the structure of the market economy is focused on individual action as opposed to larger food-system change. While other interventions in the food system, such as CSAs and co-ops, push for the creation of an alternative food production and distribution system, SHARE occupies a distinct position focused on the role of consumers within a larger market system. To SHARE, the market economy can be made to operate for the needs of low-income people if individuals purchase food collectively.

SoS—Seeds of Success

SoS is focused on transitioning individual crewmembers out of poverty and preparing them for fuller participation in the job market. While SHARE is narrowly focused on encouraging greater participation in the market economy, and CHUM and RFP intensively focused on food access and stabilization, SoS—as part of Community Action—is grounded in the theory of social development, which is based on asset development, job training, capacity building, and self-determination in the community (McKnight and Kretzmann 1993). As noted in Chapter 4, CAD's repugnance to making poverty manageable has translated programmatically into a panoply of projects knitted together under the auspices of development. As Gray (2010) notes concerning social development organizations, their "interventions fit well with the neoliberal notion of positive, productive or active citizenship where 'independent contributors' i.e. economically active people, are held as the example against passive, dependent welfare recipients" (468). Similarly, the social development model demands costly and labor-intensive interaction with participants, which is unfeasible under neoliberal social conditions and efforts to make the interventions less expensive and risks them becoming perfunctory and non-transformative. The tripartite mission of SoS of improving job training, improving food access, and neighborhood revitalization is tilted towards job training and organizers argue that although they are an urban farming program, SoS is not focused on food systems change. In contrast to the other programs, it is much more focused on working with people to ensure that they learn job skills such as showing up on time, completing the task assigned, and illustrating to employers that they are ready to work. Part of their mission is to help crewmembers understand the rules that govern the state and civil society institutions in order to empower them to use these instructions to their advantage. In contrast to traditional alternative economy programs that help participants rethink the traditional capitalist economy, SoS works to explain and define the existing economy.

In the same way that SHARE is very well defined as to their mission and focus as an organization, SoS attacks the problem of preparing people for the job market in strategic and clear ways. Rebecca, one of the directors of the program, for example, describes how participants are taught about non-urban agricultural practices during their period of employment:

> They're really integrated into our programming here. They have financial and employment coaches. They have paid time during the week, it's two hours, where they are in a group setting with a coach from here and they talk about the things that relate to finances and employment that interest them: like if it's resumes or credit reports or "How do I become a homeowner?" all those kind of things. The goal for them is when they're done with their three months stint here, then somebody works with them to get their next job. It fits in perfectly because a big part what we do here is help people to increase their income and their assets and their financial stability.

The CAD and SoS model is to teach participants the skills they need to survive within the existing neoliberal economy. Mary Beth Pudup (2008) in her analysis of urban gardening programs examines this same sense of personal engineering in the process of teaching urban agriculture. The difficulty is that urban gardening is both a specific project, as well as a piece of the larger alternative food movement which, with its focus on individual consumer action and responsible eating, has been critiqued for its reification of neoliberal subjectivities (Harris 2009).

In order to help people transition to self-sufficiency SoS crewmembers are screened and encouraged to participate in the social safety net programs that they qualify for. In addition, they employ a SNAP outreach worker, tax preparation specialists—in order to help people take advantage of the Earned Income Tax Credit (EITC) that must be claimed when participants submit their taxes—and a "navigator" to help people sign up for the Affordable Care Act (ACA). As Paul described the relationship between SoS and the CHUM Food Shelf:

> We had in 2011 donated to them. I think they're extraordinarily necessary. All our transitional employees tend to be pretty familiar with these various options and have strong personal opinions about which one is best and what's the best return on your investment of time and all that. They're obviously doing quantitatively a lot more volume than we are. I have no illusions that we're very interesting because of our sort of hybrid market-based approach, but our volume is relatively small compared to what they're doing at present. They're a good option and they're necessary for a lot of people.

CAD staff understands the complex network of state and civil society food programs that participants take part in, and view themselves as a small niche within this web.

A central tension of SoS is that they are using urban agriculture to make people job ready, even though jobs for trained farm workers are not available. Samuel, for example, described his position in the job market as precarious when he first came to SoS:

> I have limited job history. All of my jobs have kind of been underneath the table or meaningless jobs, so I needed something current, relevant to put on my resume. So I went to the job counselor and she told me "Well, this is what I can do for you. I can help you find jobs; I can coach you through the whole resume," because it's been a while since I've made a resume.

If the goal of the SoS was to produce large amounts of food, it would be more efficient to find highly skilled workers with an expertise in urban agriculture. However, as Paul notes, the particular goal of job training is accomplished through urban agriculture:

> Realistically, I don't think I could say with a straight face that I would recommend a job in local food production to someone who's economically marginal at this point. There's not like a community and that's where the attention is. The technical skills that people get from us are much more applicable to home gardening in a practical sense than to putting on a resume, although we do a lot of construction kind of stuff. You're not going to get an engineering license because you worked with us, but you will be comfortable using a drill.

By acknowledging the limited applicability of urban agriculture, SoS makes clear that their goal is job placement and personal change for individuals in need of better attachment as opposed to food systems change.

Creating policies and growing in administrative complexity means that food programs need to put rules and regulations into place. These rules and regulations in turn demand the attention of volunteers and program managers to enforce and institutionalize, and in the long run work against strong relationships between volunteers and program participants. In CHUM's case, because of their position serving the most marginalized members of society and institutional role as the food provider of last resort, they are in the contradictory place of reaching the most vulnerable but being forced to keep them at arm's length. This is further complicated by the relationship between community and social change. As noted in Chapter 3, SoS worked hard to create strong relationships between crewmembers and SoS staff, which were used to get them "job ready."

The Precarious and Porous Continuum of State Care

While we tend to think of the state as downsizing, for people experiencing high levels of food insecurity their interactions with the state and civil society social services programs are complex and multilayered. Table 5.1 illustrates the levels of

Table 5.1 Characteristics of Food-insecure Participants in CHUM, Ruby's Food Pantry, and SHARE.

Utilize Government Food Support	85%
Utilize Government Economic Support	65%
Utilize Government Medical Care	67%
Children in Household	42%

usage of government food and health insurance programs for food-insecure participants of food programs. In addition to food support programs, many low- and no-income people receive subsidized health benefits and rely on WIC for supplemental nutrition for their children. In addition, they may also be on the waiting list for subsidized daycare and housing. Others utilize benefits such as those from the Veteran's Administration, Section 8 housing vouchers, and General Assistance (which provides around $200 a month for adults unable to work). In order to receive benefits from the state, individuals must invest a considerable amount of time and effort into signing up and maintaining eligibility for benefits. In addition, they may also utilize the different community-based food programs in this study and other food programs in Duluth in order to make ends meet. Fyfe (2005), notes that the growth of the "third sector" has been accompanied by a concurrent bureaucratization and professionalization of this space, and thus there is a limited ability of this sector to provide any of the supposed benefits of community-based organizations such as small-size, specific knowledge of the service population or advocacy on the part of their clientele. Further, understanding and coordinating the different requirements made by these programs demands that people experiencing food insecurity have a sophisticated understanding of government and non-profit bureaucracy, as well as the tenacity to ensure that they receive each of the benefits to which they are entitled. Interaction with the state is a time-consuming skill that is honed through necessity, especially for those without full-time employment.

CHUM works closely with a variety of other local government and non-profit agencies to provide a "continuum of care" for the homeless population of Duluth. The purpose of this consortium is to identify the homeless population and work comprehensively to meet their health, housing, and employment needs. The idea of comprehensive identification and management of the homeless population is an apt metaphor for the continuum of government and non-profit programs that people experiencing food insecurity come in contact with. Overall, this continuum of care offers benefits, but they are precarious and porous in nature, forcing the individual benefit receiver to navigate the twisted alleyways of benefits carefully in order to ensure survival.

Because the state plays such a large role in the lives of people experiencing food insecurity, a key part of their actions to provide for themselves is ensuring that they understand how this continuum of care operates to ensure their successful usage of the variety of food-support programs. As Carney (2012) describes, tension in "everyday realities encountered by low-income urban populations in the US are bundled with other unfavorable social, economic, environmental, and political circumstances that contribute to an overall feeling of helplessness and vulnerability" (197). Hettie, a long-time RFP participant, explained how the distribution's timing during the third week of the month helps fill the gaps between her SNAP and disability benefits:

> I would get my food stamps in the beginning of the month and I was getting money from a long-term disability fund: I would get that the first week of the month and by the third week you are eating canned stuff, and you are lucky if you have a few eggs.
>
> [RFP is] very, very, helpful to fill in those voids because I get a lot of canned stuff and I buy frozen stuff but it just fills in... especially the chicken and potatoes are things that you have to have if you really cook at all. You cannot cook anything without an onion. Those are things that I don't have to worry about because I know I am going to get them here and it has also given me the dignity of not having to go to my brother and saying "I need food this month, can you come over and can you feed me tonight?" or anything like that.

RFP benefits provide one portion of Hettie's fragile survival strategy that consists of both government and non-profit components.

Phillip, who arrived at SoS out of a period of homelessness, is also very dependent on the interaction between different state programs. CAD helped him to sign up for a range of benefits that allowed him to find a stable place to live. One portion of this transition for him was establishing a "groove" of filling out paperwork and proving to various state agencies that he was eligible for their programs. However, this is not a simple process and Phillip experienced rejections and miscommunications about what paperwork was needed to qualify for which programs:

> Every six months I have to get a review but now I am in the groove, I know what they want, I already know to have [everything] all ready and pretty much know what they are going to ask for. I have got that done. Since I got Medicare that kind of loops in everything. So I fill up the application for any medical services and then I got a letter saying I don't qualify for MNsure and then I had to fill it out again for just MinnesotaCare and I had questions about it and I called, and I called, and I called, and I called, and three days later I said forget it, I cannot even do it. It says if you are on social security you don't have to send proof of residence. Then there comes a phone call and they say you didn't send your driver's license. So I showed my driver's license and put it in there because that is what they are going to do to you. It's like I have been dealing with this for so long and it is guaranteed every six months but it seems like about every other month I have to do something whether it is Medicare, Social Security, food stamps, Medical Assistance, whatever it is. It seems like I am always doing something.

Learning to navigate these waters is a new life skill that Phillip had to learn in order to move from homelessness into secure housing.

The dependency on government programs meant that recipients were constantly concerned that their eligibility would be revoked thus forcing them to rely more on other networks. The continuum was fragile because mistakes in paperwork or small changes in program requirements could substantially alter people's quality of life. Marisa, a 78-year-old white RFP participant described the nervousness of going in to get the paperwork approved for her government assistance:

> They used to do my reviews once a year over the phone. I just had to go in this time. That was the first time probably in a couple years that I had to go into the office. It's usually over the phone for me and I really like that. Going in is always ... I just get nauseous even just, because I'm so nervous about ... I don't know what it is. You don't want to hear the word, "No," or, "You can't receive that," so it can be a little scary.

There is constant contact between the social workers administering SNAP and other benefit programs and food-insecure individuals. These interventions are potential times when stigma and class bias can be communicated. For some, these dealings were time-consuming but nothing more. As Samantha from CHUM said, "It's normal everybody needs help." Or Debbie from SoS, "It's the best thing ever. It can help with paying for food. Who wouldn't want that?" But for others these interactions were places where judgment was placed on them by outsiders. Alyssa from RFP talked about the indignity of being told by a WIC official at what age she should stop her child from using a bottle. While the overall program is to her liking, she felt that staff talked to her in a derogatory way:

> One lady, I don't know if it was just a bad day for me or what, but she asked me if my little one is drinking out of a bottle and he wasn't even one yet. I said, "Well yeah, he's 11 months old." She's just like, "Well, you know that they need to stop drinking at one year of age?" And I was just like, "All of my kids have stopped drinking at one year of age. At their birthday, they get no more bottles." That's just the way it is. And they're fine with it. I was just like, wow, you are so judgmental.

Food programs necessarily interact with the most intimate parts of people's lives, and community-based food programs carefully calibrate how they interact with participants. In this sense, "performing" for case managers becomes a key part of the ability to receive state benefits (Pollack 2009), and the state oversight of the actions of individuals repositions itself from an exterior to interior project. Carney notes that many of the low-income women in Santa Barbara who would benefit from SNAP benefits do not receive them, costing the local business community enormous amounts of money (2012). Shannon from RFP notes the tension in using these benefits because of the stigma associated with using SNAP and WIC:

Like so many things, it really played a vital role in being able to get food and keeping my kids and myself fed. At the same time, there was always stigma attached to it. From the case manager or income maintenance people. So you go into the Government Services building and you fill out this form and the person on the other desk is not nice and rude. Not always, but often and actually outright cruel a lot of times.

I've had that experience. And then you go to the grocery store and with your food stamps, you buy soda and chips, people look at you and glare at you. But if you buy fruits and vegetables, they're pissed off at you because you're buying things that they can't afford. And so, no matter what you do, there's always this "hmmph."

As a researcher studying food access programs in Duluth, I became very knowledgeable about each of the different food programs available to people, but the only other people who had cataloged the different programs out there were program participants. Program organizers tended to understand the particular rules and requirements of their program, but not how other programs operated. Participants who migrated between the different programs were intelligent and strategic with their use of community-based food programs. Participants chose which program to utilize based on issues such as cost, time, and food quality. Rachel, a 58-year-old white SHARE participant compared SHARE to another short-lived local program called Angel Foods Ministry, which offered discounted food but without the flexibility that SHARE offered. They sold boxes of food that were designed with the types of food specific demographic groups, such as families with kids or individuals living alone, might eat. She noted:

Sometimes some of the things are more expensive or comparable to what you get at the grocery store. John's our grocery shopper primarily, so he looks at the list and if it's something that we can use before it goes bad or if it's a good price, then we'll get it. But we haven't done that often. I don't think he liked waiting. John never likes waiting. I think that's what it was.

Similarly, a 52-year-old white CHUM food shelf participant preferred to use the food shelf at another local church, the Vineyard, which had a reputation for offering more food than the other food shelves and made more meat available. However, because of their location away from local bus lines, they are more inaccessible than other food shelves. He said:

I liked the Vineyard the best. I just don't go there that often, because it is a long ways [away]. You have to walk six blocks from Kenwood. I like what they give you. They give you several different things of meats, usually. They are pretty good people.

In the same way that there was a rift between CHUM and RFP over their competition for donations from local businesses, CHUM shared participant lists with most other local food shelves but not the Vineyard.

Participants understood that the local food shelves were coordinating with one another and should be understood as a coordinated system of food access and not individual institutions. Sampson commented on how this coordination made it difficult for those who needed more support from the food shelf to get those needs met:

> Well, that there, when you mentioned something about the negative side of things, that's one of them, because you can only use this food shelf. If you go to any other food shelf you can't use this one. I strongly believe that if a person is in dire need of food, they should be able to use as many food shelves, take advantage of the programs that they have out here because it could be like, okay, you got some places out here that serve food and all of that, but what if you're not able to? What if you get sick? What if the weather stops people from getting out the doors to come to eat? I'm looking at that so, it's like ants, gather as much food as you can before stormy or snowy weather hits so that if you want to go outside or if the snow is all neck deep, okay, I'm not worried, I've got food here, I can cook it up myself.

CHUM helps participants sign up for the services of SNAP benefits, WIC, encourages people to use Ruby's Food Pantry, and had preliminary conversations about starting a SHARE site in order to serve the needs of food shelf participants through a more long-term program. There is interaction between community-based food programs, but there is not a comprehensive system of coordination between them to serve the needs of those experiencing food insecurity. Therefore, it is incumbent on program participants to know the benefits and drawbacks of each program and how they can be used to create a fragile form of food security. In the same way that participants have sophisticated understandings of how to preserve and save their food, they also understand the benefits and drawbacks of using different food-assistance programs.

Because each food access program attacks food security in a small and particular way, the comprehensive bank of knowledge of the rules and regulations of each program that participants built for themselves was designed to help them navigate through the continuum of care, not break free from it. Participants came to understand the specific values of food that each program constructed, and devised avenues to most efficiently feed and nurture themselves. For example, many CHUM participants would turn down the milk and eggs they were entitled to because they were too heavy to carry on the bus, and could be purchased more cheaply at local convenience stores. Others would try new programs to make ends meet, only to find out that their needs could be met better by other programs. Kim, for example, criticized the location and inaccessibility of RFP:

> I went there twice. I gave up. I went there twice on a snowy day. They left me in an empty parking lot. I never went back. Once or twice, that's it. I give up. I found out that they go once a month. I'm not willing to take that chance again. It's too long of a bus ride. About an hour and a half and then coming back. It's too much. I've only been there once. […] I don't think I'll try going up there because it's too much.

As Cindy—a CHUM volunteer—explained, the role of volunteers was also to make sure that participants knew about the variety of programs available for them:

> Yeah, people know where things are and they know where to go, when they have to be there, and, for instance this thing is like the free school meals. We encourage people to be aware that there is breakfast and lunch at school and that they need to get their kids into those programs and, you know, any way that we can to help them stretch their food and to ensure that their kids will be able to learn because they have got a decent meal and the whole welfare system is a full-time job, you know.

In the same way that working at the food shelf precluded more systemic food activism by volunteers, navigating the continuum of care was a full-time job for participants.

One aspect of the work of many community-based food programs is pushing to increase the ease in which state benefit programs are administered and can be used. SHARE and the SoS farmers market each accept SNAP, and CHUM makes sure to screen people for their eligibility for all state programs, especially SNAP that will directly address that access to food. The difficulty of filling out the eligibility paperwork and the low level of benefits that many applicants qualify for complicate this process. Dahlia, the SNAP outreach coordinator for CAD, described the dilemma in this way:

> The vast majority would only be eligible for that $16 minimum benefit. With the application now being 16 pages… ten of which [the participant must] fill out. There is a new 60 [years of age and older] application that only requires two pages of very, very basic information and then some supporting documentation. You no longer have to submit social security verification. They'll look it up for you. That is, hopefully, going to increase the number of SNAP participants.

By making the paperwork easier and encouraging people to fill out the application, Dahlia hopes to increase the number of people receiving state benefits. Winne (2008), for example, argues for the creation of a simple national form for food aid, as opposed to the hodgepodge of state-administered systems. However, because benefit levels are so low, his work will increase food support usage but not sizably alter their level of food security. Rollinson (1998) uses the term "shelterization" to refer to the process by which homeless people become regularized into the process of living in a shelter, as opposed to transitioned out of the shelter.

The patchwork of government and civil society programs that people experiencing food insecurity use create a porous continuum of care that is filled with paperwork, lines, stigma, and all too often only a marginally improved level of food security. For those engaged in this process of moving from free distribution space to free distribution space, the city and its participant institutions become a space of isolation as opposed to general inclusion in the social life of the city. As Hettie, a RFP participant, put it:

> There are so many people out there, I cannot tell you how many people that work for a living and they have kids and it's not just enough that they cannot get any aid. It will be better for them to be jobless. They will be better off to quit their job and stay at home because of the facilities and they would have aid, but instead you choose to go to work every day and stick by and lot of this is in the city, but still that shouldn't happen, but it is frustrating and it is more frustrating now that I am on this. It was frustrating before, now it is more frustrating.

CHUM user Kim, a 50-year-old Native American unemployed woman argues that her key goal is not to use the food shelf but instead to find a job:

> Finding a job, I would rather work. I have been waiting now for how long has it been, the other government services don't find me a job and that is like rehabilitation. First, they had me go to a psychologist, my drinking kind of messed up my memory. So I went through that and they found out I wanted to stop and then it is slow, and I am anxious to get to work.

Because of the high percentage of CHUM participants who use state benefits, this was a common sentiment: participants wanted to be in a more secure financial position but needed state services to survive. Volunteers and participants were therefore in a holding pattern that consists of enumerating the failures of the current food system by recording how much food comes in, carefully allocating enough food for individual families. They were also doing the heavy emotional work of providing for people in difficult situations amid the reality that long-term change was not coming and their position on the frontline was likely permanent.

The Disciplinary Project of Food Access

Community-based food programs have a complicated relationship with the neoliberal state. The state creates the macro-level conditions that produce food insecurity as a particular manifestation of poverty and economic inequality. CHUM, RFP, SoS, and SHARE are situated interventions into the problem of food insecurity, but, in the same way that they only articulate alternative economies in fits and starts, they do not have a language to discuss the role of the state in creating the overarching problem that they are solving. Instead, they carefully fulfill their specific mission while enacting a different relationship between the food and community.

At each of the community-based food programs, narratives and policies that support workfarism and neoliberalism are present. At CHUM, there is a regime of numeracy wherein individuals are carefully counted and allocated aid according to preexisting formulas that create distance between program participants and volunteers. This process inspires participants to police each other and works to further erode a sense of commonality among food shelf users. Similarly, at RFP, this takes the form of "pious neoliberalism" (Atia 2012) within which faith and self-sufficiency are tied together under the same rubric of proper behavior. SoS engages in this project through an embrace of social development that clearly defines the market economy as the preferred direction for individual program participants. Last, SHARE's program design supports entering the market on terms negotiated by a larger collective effort but not a calculated dismissal of the market's power. These messages become more pronounced because they exist within a frayed continuum of care within which each state agency or civil society program has rules and regulations that people experiencing food insecurity are forced to follow. Because alternative economies are not forcefully articulated and promoted, the wider discourse of neoliberalism prevails and becomes the norm.

SoS and CHUM are unique in that they make sure that all participants are receiving the government aid for which they are entitled. Although this is not framed as a transgressive project, it is important because it recognizes the relationship between hunger and state policy and puts pressure on the state to respond to documented need. Equally as important, the steady stream of aid which SNAP benefits bring, and the health benefits that come with enrollment in Medicaid brings, supports people in precarious situations. This pedagogical project of teaching people experiencing food insecurity to be strategic users of the state recognizes the time-consuming work that individuals experiencing food insecurity often perform in order to survive. This, combined with the labor necessary to procure and prepare food comprises a distinct set of local knowledge created by individuals in a marginalized position, which reflects their ingenuity and survival skill. The neoliberal state that community-based food programs push up against work together to create a porous continuum of care that assists those at the margin to survive but not thrive.

This means that for many marginalized members of the community, existing in Duluth is possible and services are not viewed negatively but are instead supported. Many very poor users of the CHUM Food Shelf admired the multiple ways in which they could feed themselves in Duluth and saw the Duluth food access community as special. As Alan, a 47-year-old African-American formerly homeless CHUM participant described it:

> No, our system here is good. I have nothing to say about Duluth and the way they help people here. Even the homeless, they always say "There is no reason why you should be hungry in Duluth because you have all these places

that feed." All you got to do is show up, and you're not going to get turned down.

However, in order to create this porous continuum of care, community-based food programs must create and enforce guidelines and force people experiencing food insecurity to maneuver between different programs in order to find the services that they need. Enforcing guidelines, as explored in Chapter 3, works against forming community between participants and volunteers, and between participants, as they see inequitable distribution of the scarce commodity of food. Government programs become a meta-framing for how people view their relationship to the world.

SoS and SHARE are unique in their commitment to teach participants how to be effective market citizens. SoS pays for their crewmembers to attend financial literacy programs, and SHARE volunteers try to teach participants how collective purchasing works and why SHARE is able to help them without relying on surplus commodity distribution. This movement away from the charity distribution model meant that the food they distributed was of higher quality, and participants had more control over what they took home with them. This suffused the programs with a critique of charity and embodied more of a sense of dignity and empowerment than the assortment of different food products available at CHUM and RFP. By choosing to empower participants and treat them as economic citizens, SoS and SHARE engage strategically in the neoliberal project of defining citizenship by market participation, with the goal of using the market as a pedagogical tool to address marginality.

Any attempt by community-based food programs to articulate a message distinct from neoliberalism and workfarism is complicated because the dominant message presented by this porous continuum of care is that individuals are powerless, a burden on society, and in need of a modicum of support in order to survive. Whiting and Ward, noting the high stress levels associated with poverty and low food security note, "At the very least, these results suggest that food programs serving reservations communities need to find ways to limit the kinds of procedural obstacles that procures high levels of stress among program participants who already face the multiple challenges presented by living with poverty and food insecurity" (503).

In the final chapter, I turn towards the relationship between food access and the numerous different agencies, organizations, churches, and groups that operate in the shadow state. I argue that given the prevalence of these institutions, it is important for the livelihood of people experiencing food insecurity to creatively theorize how these spaces can be used to meet immediate needs while also engaging in activism. This work demands a clear articulation of what forces are impeding food justice, as well as a pedagogical component that helps participants learn the skills necessary to maneuver in the difficult spaces created by neoliberal capitalism. Here, I examine both the larger lessons that these case studies represent for the voluntary sector and argue for a more strategic use of this interdictory space.

Note

1 From *The Conquest of Bread* by Peter Kropotkin, page 76.

6 Looking for Paths to Food Access and Solidarity

> *[Food insecurity] is an economic issue that gets framed as a food issue, because it's easier to talk about food than it is to talk about economic justice, I think.*
>
> Rebecca, Seeds of Success Organizer

> *Let's manage the things that we have to manage out of the belief in people and the belief in community. If you really believe, you don't have to hold it tight. It's like a handful of pudding: if you squeeze too tight it runs out, but if you hold it loose it stays together.*
>
> Dana, SHARE Organizer

The 14% of the US population that routinely experience food insecurity represent clear proof of the fact that the material overabundance of food does not forestall scarcity. Programs that seek to address food insecurity exist in a complex and paradoxical space; although hunger is an immanently solvable problem, anti-hunger organizations are an unquestioned part of our cultural landscape and often do little to empower the hungry or transition those experiencing food insecurity to a more sustainable place. Worse, by taking care of the poorest members of society they fill in for the "rolled back" state and turn their clients into disempowered participants, and thus work against the already difficult project of unifying and politicizing this disparate group. What are the possibilities of more supportive and nurturing projects to "roll out" of the voluntary sector and better support this population? When asked what to do about the problem of hunger, Poppendieck (1999) sees widespread food insecurity as a problem that has emerged because of a fundamental breakdown in US society. She argues:

> When I say that the fundamental problem is growing inequality, I do not just mean that the poor are further from the rich or that they are less well off in absolute terms, and thus more likely to be in need of emergency assistance, although both are true. I mean that our society is coming apart that we are losing the sense that we are interdependent, all part of the same society.
>
> (306)

In examining the work of RFP, SHARE, SoS, and the CHUM Food Shelf each organization creates an alternative economy with a particular focus on improving food access and they each work with varying degrees of success to fight community disengagement and marginalization. Because of the rollback of the state in the 1980s, the anti-hunger movement has been focused on food provisioning and forestalling the immediate effects of hunger. Transitioning to a more comprehensive movement based on food security demands changing from a client/service provider model to one that works with people to help them push both the state and existing distribution networks to better provide them with the food they need.

This process is both complicated and aided by the overabundance of commodity production in the US. Even though billions of dollars are spent on commodity subsidies, crop insurance, and state food benefits for the poor, these programs are not meeting the needs of the growing population of low-income Americans. This glut of food production creates scarcity and thus sets in motion the demand for redistributive polices. However, this surfeit also creates a visible disjuncture between the presence of scarcity and the reality of food abundance that in turn produces demand for more equitable distribution. The most revealing moment of RFP, therefore, comes when program users ask, "How could there be so much donated food?" Now this question is routinely answered within the model of charity and generosity, instead of an insistence that this abundance is a signal that food should be distributed more equitably.

An unresolved tension in the case studies is how to solve the immediate needs of people experiencing food insecurity while also creating a space for people to fight for an end to poverty. Michelle Alexander (2012) argues in her critique of the mass incarceration of African-American men that there is not a tension between attending to the needs of the imprisoned *and* advocating for a different system. She argues "reform work *is* the work of movement building; provided that it is done consciously *as* movement building" (235). And this consciousness is what is missing: those in the alternative food movement are not consciously enough considering the needs of those experiencing food insecurity, and those serving the hungry seem not to be engaged in rebuilding the food system (Allen 2004). As Mark Winne (2008) writes, "the cost of healthy food, which should include local and organic whenever practicable, should not be a limitation for any class of citizens" (186). Working to ensure all classes of citizens can afford fresh food demands food movement building that attacks poverty and powerlessness. SoS has addressed this issue by securing for their small number of crewmembers the state benefits for which they qualify and through working with them to explain how the economy works and how they can improve their position in the job market. SHARE, CHUM, and RFP—which touch much larger numbers of people—have focused mainly on developing a sustainable commodity distribution network to improve people's food security. However, Dana, who worked with SHARE as they went out of business (perhaps to rise again in a different form), notes that they focus on solving the needs of their participants whatever these needs may be, thus casting SHARE as a political project, effective so long as it operates as a living organism that responds to the needs of its participants:

> *Now* [SHARE] fits the bill quite nicely. It's complex, it has the same kind of social things for me, it has the service thing for me, but it doesn't have to be unloading food off a truck and kicking it out to people once a month. It can be something else if it evolves into that, because we really ... I run it like business, and SHARE, if it is to be consistent, it needs to be run like a business.

By tying program outcomes to the wants and needs of participants, organizations can stay relevant and important in their participant's lives, even as they change and evolve over time. However, as the closure of SHARE illustrates, this is difficult because most volunteers sign up to take part in a recognizable and repeatable program of distribution, not the amorphous political project that Dana recommends. I take Dana's reference to running SHARE "like a business" as a further symbol of the connection between empowerment and neoliberalism (Gray 2012), but also as a call to food access programs to be reflexive and responsive to the needs of low-income people, even if that disrupts the everyday actions of volunteering (see for example Dyer et al. 2008). For people in need of empowerment, change is more important than consistency.

In this final chapter, I look back at the key arguments of the book and offer six starting points for how organizations in the shadow state can more strongly tend to the day-to-day needs of those experiencing food insecurity in ways that fight back against clientelization and workfarism, while simultaneously operating as "spaces of hope" within a landscape that often seems bereft of empowering possibilities:

- Support community as a way to empower those in need
- Challenge neoliberal subjectivities
- Create opportunities for political engagement
- Discourage clientelism through "weak theory" and innovative program design
- Be responsive to participants' voices
- Speak to the interconnections between poverty and food insecurity.

Support Community as a Way to Empower Those in Need

Community is created for the volunteers at SHARE, CHUM, and RFP in such a strong way that the ritualized practice of serving participants, unloading trucks, and distributing food became games that good friends could play together. Similarly, through the involvement in leadership teams, individuals at RFP and SHARE were transformed into organizers of projects far bigger than themselves that were enacted through close cooperation with people who became close friends. However, the gulf between participants and volunteers tended to work against community formation among those receiving the food. In fact, it was the very process of serving the food that was the glue that tied volunteers together, and this set of actions was something that participants were left out of. If providing a space for volunteers to "give back" to the community means that authentic relationships can only be made between those volunteering, then another model is needed. As Rose (1999) notes, community is a tricky project, and charity is a technology of community governance as volunteers come to internalize what Roger Keil (2002) labels the "everyday neoliberalism" of wealthy community members giving back and supporting those in need.

SoS developed a different model based on shunting volunteers off into "volunteer days" and keeping the traditional 9–5 workday period as a place for only crewmembers and crew leaders. Interestingly, this was done within a shadow-state institution that was very corporatized, closely connected to the state, and hardly the grassroots social change organization championed by Alinsky (1971). Michael Brown (1997), similarly, found transgressive possibilities emerging in the one-on-one personal relationships that developed in large-scale traditionally corporatized HIV+ care organizations as opposed to the more radical organizations. A benefit of SoS's model was that crewmembers could solve the problems associated with growing food together, and when they interacted with volunteers the power dynamic was completely reversed and crewmembers acted as guides and mentors, showing the volunteers how to correctly harvest and work in the gardens. While this model served only a small number of people, it illustrated a way in which participants and volunteers could work together with a different dynamic. This implies that the relationship between participants and volunteers should be regulated and planned in a way that puts the food insecure in positions of leadership and management because they most intimately understand the problems inherent in a food system that produces "cheap food" (Carolan 2013).

In addition to bringing their own needs for community, volunteers at CHUM, and to a lesser extent RFP, often brought different relationships with food than their participants. For many volunteers, food was a source of pride and identity in their lives. Consistent with lifestyles within this third food regime (McMichael 2009), they purchased foods that were local and part of boutique production and celebrated this alternative food system that was distinct from industrial food. Most importantly, food was part of their class identity (Peterson et al. 2014). This understanding of food was distinct from the much more utilitarian knowledge of food that many participants in the case studies had. While the need for food may be a cultural universal, this research suggests that it might not be the best way to build bonds between different people. Instead, severing the natural connection between food insecurity and other food-based social movements might make it easier for volunteers and participants to find commonality with another.

J.K. Gibson-Graham (2006), engaging with the idea of community in the context of economic development and the creation of community economies note that the perceived fraternity and comradeship of community is a double-edged sword. On the one hand, Robert Putnam (2000) and many other proponents of the power of community argue that these ties are strong and have the ability to solve problems in a stronger and more consequential way than state intervention. On the other hand, community, with its connection to place-based identity, promotes parochial attitudes towards development and distaste for transnational communitarianism that is at the root of the community economy ideal. Gibson-Graham, interrogating Giorgio Agamben's (1993) analysis of community, argue in favor of moving away from community based on communality. They write: "the association of community with a *being* that is already known precludes the *becoming* of a new and as-yet unthought ways of being" (85). But what would these "unthought ways of being" be based on and could they survive the day-to-day demands of being together and nurturing one another? Because volunteers at CHUM, RFP, and SHARE came to feel such strong bonds of friendship and camaraderie with one another, the model of monthly food distribution does not need to be abandoned, however, moving away from the clientelist and charity-based model of change is essential. Jennie Weiss Block and Michael Griffin (2013) in their work *In the Company of the Poor* explore the theme of accompaniment in their analysis of Paul Farmer and Gustavo Gutiérrez's liberation theology influenced work. Similar to the food programs in this study, they meet people in whatever condition they are in and enter into a relationship of unknown duration without a preset plan and with the goal of knowing and experiencing poverty with those hurt by its violence. They write:

> The practice of accompaniment is highly personal and deeply relational. Accompaniment of the lonely poor involves walking with—not behind or in front—but beside a real person on his or her own particular journey in his or her own particular place and time, at his or her own particular pace. Accompanying others in their struggles for survival does not have a beginning or end, and there is no outside plan to be imposed.

Under this model the project of working to improve food access is transformed into a concrete act of accompanying the poor, as opposed to instituting a specific project. At many of the food programs in this study, the "cure" for food insecurity had developed over time and simply needed help in order to be implemented. Acknowledging that programs change over time could create the space to focus on companionship and a form of reflexive flexibility in which programs change and evolve to meet participants' needs. Accompaniment can be seen as a way of creating the "unthought" way of being that Agamben and Gibson-Graham propose. This ethos is a strong part of Catholic social activism and liberation theology (Gutiérrez 2003) as well as the Catholic Worker Movement (Day 1963). Along these lines, accompaniment could take the form of volunteers using their social capital to press for reforms to the larger food production system.

Challenge Neoliberal Subjectivities

New York Times food columnist Mark Bittman, famed food author Michael Pollan, Union of Concerned Scientist senior scientist Ricardo Salvador, and former UN Special Rapporteur on the right to food, Olivier De Schutter, have a goal of creating a social movement around the need for a national food policy in the US that challenges the unhealthiness of the US diet and enormous subsidies for environmentally destructive farming practices. Writing in the *Washington Post* on November 11, 2014 they argue:

> How we produce and consume food has a bigger impact on Americans' well-being than any other human activity. The food industry is the largest sector of our economy; food touches everything from our health to the environment, climate change, economic inequality, and the federal budget. Yet we have no food policy — no plan or agreed-upon principles — for managing American agriculture or the food system as a whole.

While they are joined in their outrage by many parts of the alternative food movement and the anti-hunger community, food simply is not something that diverse groups of people can easily rally around because it marks differences between people, and the types of food that people want access to are informed by distinct understandings of what nourishment mean (Bell and Valentine 1997; Peterson et al. 2014; Guthman 2008). More importantly, this group of writers see themselves as empowered citizens with the ability to call for and receive change from those in power. Those experiencing food insecurity are the people most in need of a food movement and they seem unlikely to join together in this mass movement because the places in which they congregate—like soup kitchens, county offices, and food shelves—do not have a space for consciousness raising and instead are designed around creating passive recipients of aid. Shoshana Pollack (2009), studying the disempowerment of lone mothers in Canada, documents the

> extreme challenges of surviving on the minimal benefits they receive, confusing eligibility, requirements, copious amounts of paperwork, bureaucratic inefficiencies, difficulties contacting workers, degrading attitudes in welfare office, and the enormous amount of time and effort (work) required to continually justify their eligibility for social assistance.
>
> (229)

Similar to those enmeshed in the porous continuum of care that provides for those experiencing food insecurity, this group is being conditioned to obey the state, rather than participate in their own governance. The situation created by this governance regime means that they have the pressing concern of survival on their mind and food does not serve at the center of their identity in the same way that it does for foodies, chefs, and food activists.

The "rollout" project of neoliberalism is a constantly evolving mission (Peck and Tickell 2002), so much so that Macleavy and Peoples (2009) describe the growth of the US military in the age of neoliberalism as a form of "active welfare" in that the military is becoming closely interwoven with the provision of welfare services and militaristic logics of sacrifice and discipline are being transferred from the military to the administration of welfare services (see also Lutz 2002). The workfarism literature maintains that neoliberalism is violently insinuating itself into new institutions in order to discipline the population (Peck 2001; Wacquant 2009b). Given the dynamism and flexibility of this system of control, activists have a moral imperative to be involved in this process in order to push the state and shadow-state institutions in more humane directions. Problematically, the main institutions that the food insecure come into contact with are designed in the charitable mold and reinforce neoliberal subjectivities; as Peter Buffett wrote in the *New York Times* in his analysis of the Charitable-Industrial Complex:

> The rich sleep better at night, while others get just enough to keep the pot from boiling over. Nearly every time someone feels better by doing good, on the other side of the world (or street), someone else is further locked into a system that will not allow the true flourishing of his or her nature or the opportunity to live a joyful and fulfilled life.
>
> (July 13, 2013)

Two examples of projects that operate within the third sector but operate in ways that unite people around a shared need for food, decenter neoliberal subjectivities, and nurture collectivist subjectivities among participants are the Harambee Neighborhood Gardens in Milwaukee, Wisconsin (Ghose and Pettygrove 2014) and the transnational Voices of Hunger projects in Indiana and West Bengal, India (Dutta et al. 2013). Harambee Community Gardens is a program that combines aspects of CHUM and SoS in that they transformed vacant lots into community-controlled urban gardening plots, developed a food shelf (that serves community-grown produce), and initiated a youth employment and job-training program. It is a comprehensive project overseen by the All Peoples Church "a dynamic Christian community in the heart of Milwaukee." Harambee, Swahili for "pull together," was instituted through voluntary sector agencies, and exhibits the practical ambiguities of utilizing this sector; for example, many of the key agencies and volunteers are white while the neighborhood is predominantly African-American, and while volunteer labor is open to all, many working people lack the time to volunteer, and residents with disabilities have difficulties being involved. However, the project has involved neighborhood residents through community control of garden plots, which have space set aside for community recreation and celebration, and now trained youth and neighborhood residents act in a new capacity overseeing this shared neighborhood space. In this program, neoliberal citizenship subjectivities are challenged by participants who were empowered to create something under their control. In particular, All Peoples Church negotiated for control of neighborhood space with the City of Milwaukee, an example of both political empowerment and mobilization around meeting community needs collectively. Indicating the limited spatial and transgressive potential of this project, Ghose and Pettygrove argue that "these community gardens create claims to space and resist local government policies that prioritize for-profit development whether or not they concretely impact state policies" (17).

In this vein, Dutta et al. (2013) analyze the Voices of Hunger project, an academy/community co-constructed performance created as part of the Culture Centered Approach (CCA) to health disparities. The project, similar to CHUM's project of training community members to advocate for themselves in political forums and the Big View project of CAD connected the experiences of the hungry with the tools to advocate for their self-betterment. Dutta et al. (2013) define the CCA as programs that "seek(s) to address health disparities by fostering opportunities for listening to the voices of those at the margins through a variety of participatory communication methods such as co-constituting data gathering and analysis, community dialogues, PhotoVoice exhibits, community driven White-papers, community-driven media advocacy and town hall meetings" (2). By working with communities in India and the US, they illustrated the common theme of disempowerment in hunger. The project "inverted the logics of neoliberalism" by portraying food-insecure communities as hardworking, and created interactive performances highlighting the difficulty in signing up for state benefits and addressing the "everyday barriers" that marginalized people have providing for themselves and their families.

At each of the community-based food programs, organizers have a strong and complex understanding of how their work is engaged with a spiritual practice or tied to a particular understanding of social change. Taken collectively, they offered sophisticated and groundbreaking theorizations of the relationship between food, society, and poverty. For CHUM and RFP, different understandings of religious beliefs provided this glue, and at SHARE and SoS it was a more secular understanding that people could be educated to act in ways that would empower them to change their surroundings. At SHARE, this ideology motivated them to eschew the model of surplus distribution in favor of empowering participants to work together to purchase food, and at SoS the belief in not making poverty "tolerable" meant that temporary food aid was not pursued in favor of job training.

By not more fully embracing the revolutionary project of transgressive language and conscious rearticulation of the market economy, the projects in this analysis avoid an opportunity to expand the community economy endeavor into new places. The discursive power of neoliberalism and workfarism highlight the importance of using civil society as a place within which to construct counter-narratives to this totalizing project. Fortunately, these discourses already exist. For example, faith can be used to constrict a counter-narrative to the withering effects of the logics of capitalism and globalization. Bedford-Strohm (2012) writing in the tradition of liberation theology, which advocates a "preferential option for the poor" in biblical interpretation, argues that poverty should be defined as "lack of participation," and hunger should be understood as "lack of participation in multiple ways" because food insecurity is a result of social and economic exclusion (2). This theological critique of hunger grounds religious-based food activism in the critique of marginalization seen in the work of Watts and Bohle (1993), and provides a powerful discursive counterpoint to neoliberalism.

Replacing neoliberal subjectivities with empowering collectivist identities can be an embodied and joyous project. Gibson-Graham (2006) conclude their extensive analysis of the film *The Full Monty*, within which unemployed steel mill workers create a new subjectivity and position within the economy for themselves through staging and performing a strip show, by arguing:

> *The Full Monty* is, for us, about the potential for multiple productions—the making of masculinities released from the symbolic centrality of the erect phallus and manual labor; the creation of new forms of community energized by pleasure, fun, eroticism, and connections across all sorts of divides and differences; and experimentation with a communal class process in which interdependence and incompleteness are accepted as enabling aspects of individual subjectivity.
>
> (18)

The design and implementation of all of the programs in this study seem designed to thwart the sort of enthusiastic embrace of the possibilities of an economy based on abundance, which is seen in *The Full Monty*. Cindi Katz (2004), for example, in her work on play, argues that the freedom of children's play allows them to try on new identities and positionalities. Playing with the idea of entering into alternative economies and subjectivities through pushing people out of their comfort zone, Natalie Doonan (2014), in her work on the *SensoriuM* dance troop based in Toronto, analyzes the interactive performance *Botanical Animal* in which Amanda Marya White works with audience members to make and can salsa made with tomatoes whose seeds actually passed through White's digestive system. The performance art project brings audience members face-to-face with the industrial food system and forces them to confront the anomie in this system in an affective manner. Her project is in direct conversation with Gibson-Graham and efforts to oppose neoliberal subjectivities, as she notes, "Self-discipline is instilled in populations as they voluntarily regulate their bodies in accordance with marketing and statistics published in studies on health and safety" (39).[1] While globalization allows for the consumption of tomatoes out of season and independent of climate, White has aligned her own body with the growth cycle of the tomato. She thus presents "a refusal to simply adapt her own subjectivity to the demands of capital" (40). As theater of the oppressed illustrates (Boal 1996), using performance as a tool of social change is not unique to food systems, but thinking creatively about how to illustrate the power of new subjectivities in cheerful and unexpected ways is something missing from programs that serve the food insecure.

As noted in Chapter 4, food-insecure individuals are deeply enmeshed in the survival of their own families and households. Thus, expanding food access programs in ways that recognize the importance of these relationships and strengthen these connections reframes individuals from people in need to members of households seeking to strengthen and improve the well-being of their familial community. For example, organizations could follow RFP's model of providing extra food for participants to share with their extended family, provide safe and secure play areas for children so all family members are welcomed at distribution sites, and ask after the food security of all family members as ways of valorizing the variety of community commitments that food-insecure people have.

Create Opportunities for Political Engagement

The federal government plays a large role in providing food support for those in need. Fully 85% of the food-insecure individuals in this study used some form of government food support. In addition to TEFAP, SNAP, WIC, and the other USDA-managed food support programs, the federal government also supplies direct subsidies to commodity crop production, which, according to data collected by the Environmental Working Group (2013), amounted to just under $300 billion between 1995 and 2012. Many in the US believe that a government that governs least is best and prefer to look for civil society responses to food insecurity. However, the important role the federal government plays in providing for people experiencing food insecurity suggests that those interested in changing the food system should engage with federal policy and not focus their efforts on community gardens and food shelves.

CHUM and SoS each made a concerted effort to sign people up for the state programs for which they were eligible, and constructed these benefits as a more sustainable solution to food insecurity than relying on civil society programs. Although SNAP usage is positively correlated with food insecurity and often last for less than the entire month (Nord and Golla 2009), these benefits become a baseline upon which other purchased food or food from other community-based food programs can be added in order to make ends meet. Low-income people are deeply intermeshed in the complex world of applying for and maintaining state benefits, and only food-insecure people themselves understand how the various elements of the porous continuum of care knot together. In fact, as I learned more about the state programs and alternative ways in which to access food, then through interviews, I began educating social workers and food shelf workers about how this system worked. Because people experiencing food insecurity have such specialized knowledge about how the process of rollout neoliberalism works, they seemed to be well suited for political activity: they understood the drawbacks and benefits in each of the programs and had ideas about how those programs could be improved. Their knowledge about how to secure state benefits could be transformed into the ability to engage in more sustained discussions about how the entire food system could work better. Advocating for stronger state programs creates agency among the marginalized and highlights the important role of the state in creating surplus and maintaining scarcity.

RFP and SHARE had a more ambivalent relationship with the state and structured their programs in ways that highlighted the importance of non-state actors. RFP saw themselves as a "hand up not a hand out" and wanted to provide food to moderately food-insecure households in order to help prevent their reliance on government programs. In contrast, SHARE grew out of the activity of a set of activists promoting community empowerment through collective engagement with the market. To them, there was power in the creation of a volunteer-led network that met the needs of the poor without the expertise or authority of government. For each of these groups, their focus on civil society responses to hunger took attention away from the ways in which the neoliberal state transfers wealth from the poor to the rich and takes away from the social safety net. This research suggests that such a stance continues the project of delinking poor support from the state and positioning civil society as the legitimate space for the delivery of welfare services (Macleavy 2009).

Watts and Bohl (1993) argue that food access and hunger are symptoms of political disenfranchisement, and of the case studies only SoS through the Big View program and CHUM through community-organizing efforts engage in actions that promote the politicization of the hungry. None of the programs promote civil disobedience or direct radical action against the state. As Nik Heynen (2010) writes of the public theater of FNB:

> Unlike much charity, FNB works hard not to be complicit in the perpetuation of the capitalist states' biopolitics, but seeks to radically transform it. Unlike much charity, and because it is a movement of resistance, rather than amelioration, FNB might just threaten the developing, privatised modes of regulating the poor. FNB does this—in, however, minor a way—by cooking up not charity but mutual aid.
>
> (1233)

The groups in this study are so focused on "amelioration" than radical theater that questions the roots of the problem are not engaged in. The next frontier of food activism is projects that question the roots of surplus, marginalization, and hunger in a way that also nourishes and provides sustenance to the questioners.

However, this type of direct activism may not be possible. A problem with radical activism within voluntary sector efforts around food access is that these programs must be framed and structured as permanent and reliable pieces of the quotidian lives of people experiencing food insecurity. Periodic discounts in food prices, salvaged foods, and spectacular protests may document outrage, but they do not engage in the ongoing process of the day in and day out provision of nourishment. Andy Merrifield, in *The Politics of the Encounter* (2013), argues that protest is at its core outcry as opposed to mobilization. He argues, "The crowd at a mass demonstration, a mass encounter, or a mass occupation expresses political ambitions before the political movement necessary to realize them is created" (97). He continues by using Occupy as an example of what he calls "minor space" imbued with "negative capability" in that it clearly enunciates the problems with the present system without advocating for an alternative. Food activists must build and construct on a daily basis, which perhaps weakens their ability to mobilize against power. Instead, food security must be understood as resulting from a confluence of forces coming together in households, which allows them to have sufficient funds to purchase the food they desire and sufficient political capital to lobby for food policy that supports their lifestyle.

One way to address the problems with USDA programs is to encourage greater involvement in the political process among people experiencing food insecurity. Food access programs tend to position themselves as non-political social service programs which are far from the activist/social mobilization and power organizations that seek to empower communities and encourage greater political power. While no data exists that clearly analyzes the relationship between voter participation and food insecurity in the US, low-income people, younger people, and Latinos tend to have lower participation rates than the population overall, and high-income and well-educated people tend to vote in much higher numbers. Registering people to vote at food pantries as the CHUM Food Shelf does help to work against this trend, and should be a part of all food support programs. Alinsky, for example, describes his mission as one of empowering citizens in industrial areas in order to "break down the feeling on the part of our people that they are social automatons with no stake in the future, rather than human beings in possession of all the responsibility, strength, and human dignity which constitute the heritage of free citizens in a democracy" (109, quoted in Horwitt 1992). The action of political participation works against the positionality of a disempowered automaton while pushing for government support.

Because people experiencing food insecurity are shrewd utilizers of state programs, community-based food programs could be critical spaces where participants are empowered to push for better and more appropriately designed state programs. Right now the multiple institutions of the neoliberal state interact to create a porous continuum of care and a geography of isolation within which the specific rules and regulations of each community-based food program are more important than acquiring the tools to transition out of this space. While the specific rules at each program are not too difficult for participants to solve, they become one more set of rules that limit the participants' ability to engage in actions that might be able to help them transition out of poverty. Using community-based food programs to train participants to push for better-designed benefit programs would change the culture of rule teaching and following and provoke more empowering discussions about how rules are constructed and enforced.

A benefit of politicizing community-based food programs is that it would teach skills to people experiencing food insecurity, which could alter their material position and create a common bond between volunteers and participants. As noted in Chapter 4, participants at all four community-based food programs expressed interest in meeting politicians and engaging in other community work so this change would not be moving away from the desires of participants. Although volunteers at CHUM and RFP often had well-developed critiques of the food system, their particular work distributing food only began to enact their transformative dreams. Similarly, although people experiencing food insecurity had sophisticated and systematic understandings of how food access programs worked, they did not know a lot of the process of political lobbying and political organizing. Leonard, a 52-year-old white unemployed CHUM user, expressed his feelings about political engagement:

> I would be willing to maybe get involved in something, filling out forms or doing something. I don't know, to help us get more food or something of that nature. I don't know exactly what I could do.

The structure of food support programs that fund commodity crop production as compared to "poor relief" programs like WIC and SNAP is especially problematic. In contrast, commodity crop production is supported by direct economic subsidies that operate through traditional Keynesian demand-side economic development in rural areas, which create a ripple effect and help support universities, restaurants, and farm machinery companies, which support the industrial farming model. In sum, the USDA creates jobs in rural communities and pushes urbanites into a labor market where their skills are not valued.

Designing food access programs to address both the immediate condition of hunger and the socio-political circumstances that create hunger is difficult and demands a concrete linkage between ideology and action. For example, the community organizing work at CHUM is defended as being a key part of the organization's work through interpretation of scripture in a way that promotes collective action to further the needs of the poor. At a training about Saul Alinsky (1971), one pastor cited the gospel story of Jesus forgiving and healing a paralyzed man (Mark 2:1–5). In the story, collective action is taken to help a paralyzed man reach Jesus, and Jesus rewards this camaraderie:

> Some men came, bringing to him a paralyzed man, carried by four of them. Since they could not get him to Jesus because of the crowd, they made an opening in the roof above Jesus by digging through it and then lowered the mat the man was lying on. When Jesus saw their faith, he said to the paralyzed man, "Son, your sins are forgiven."

To this end, Ballamingie and Walker (2013) explore the work of Just Food Ottawa where various members of the local food community created a community-based space where food is provided to those in need, and where the principles of ecological sustainability and co-operative decision-making are followed. The program pushes legislatively for a $100 healthy food supplement for adults on social assistance through the Put Food in the Budget campaign and is developing a garden/food hub to connect and teach urban agriculture to Ottawa residents. In this project, they "exist simultaneously in a multitude of economic movements, ranging from market to alternative market, and non-market" and in community economies (Ballamingie and Walker 2013, 539). As such, their project is always "fundamentally performative—*continually in the process of becoming*" (540) as they negotiate with the state and private funders and find a way to conduct their operation in light of their ethical principles. In light of the experiences of CHUM, RFP, SHARE, and SoS, the embrace of "performativity" and "becoming" is essential because there is not a single project or way of addressing food insecurity. However, through the praxis of developing a theoretical commitment to those in need, creating policy, and funding the work, the contours of the possible becomes written. SoS, for example, realized that growing produce *demanded* a constant rethinking of how to put food into the hands of those in need, which changed from produce boxes to a farmers market, in light of what was possible.

Discourage Clientelism Through "Weak Theory" and Innovative Program Design

Gibson-Graham (1996) argue that we should embrace the emancipatory potential of "weak theory" in contrast to the overarching narrative of capitalism and communism. As Derickson (2009) notes, "theory building, especially that with progressive political aims, must be reflexive, deliberative, and in dialogue with a diverse communities of knowers" (13). Propitiously, each of the programs in this analysis contain just these kind of "knowers" whose need for physical support and interest in working proactively for their own self-betterment contain elements to be built on in building weak theory. However, two problems exist in this process. First, as has been noted, these knowers must be brought into the process of theory construction and program design in order to help shape this knowledge production. Second, Marxism and capitalism are such powerful narratives that "pushing back" cannot be done weakly, even as the term "weak theory" implies a cautious and careful push back. This is a delicate balancing act because opposing neoliberal workfarism must be an assertive part of program design if it is to be successful.

The space that CHUM and RFP created where people of faith came together to serve participants at vulnerable times in their life created an opportunity for the reflective construction of weak theory. Deborah, for example, is a regular volunteer at the CHUM Food Shelf where volunteering with her elementary-age daughter is a weekly activity. Deborah's daughter stocked food and spent time with both volunteers and participants in a locale very different from her usual middle-class surroundings, but within the charity model her ability to use this time to invent new possibilities for relationships across class lines was not allowed to develop. Deborah described the experience in this way:

> Well, there is many a Tuesday nights that she'll come home and open the fridge and say "Oh my gosh mom, we have got way, way, lots of food, we have got too much food." Yeah, we have got a lot and then she will talk about

how lucky we are and how many of the other families that were in tonight that were happy to have both a can of pork and beans and a can of potatoes or whatever it was, they got excited about it because that would last them for the next couple of days. She is like, "Mom, we eat that down in like five minutes, we are just pigs." I was like, "Well, we do have more than a lot of the people that you met today" or whatever, but it usually comes up when we are eating or grocery shopping. She will talk about grocery shopping and comparing prices and, "Mom, look there the canned corn is 70 cents" or whatever and "that family was so excited last week when they could get two cans of corn. That's only a $1.40. Maybe we should have just given them $5 so that they could have got more corn."

Through her time at CHUM, Deborah's daughter is able to understand the complexities of food support and civil society actions in response to this need, but only able to imagine a future in which her generosity can address the issue of food insecurity.

Sara Miles (2007), in her memoir *Take This Bread*, gives an example of a food shelf that challenges the paternalistic and disciplining actions of traditional feed shelves. At St. Gregory's of Nyssa Episcopal Church in San Francisco, the food shelf operates as a corollary of their practice of open communion: in the same way that the sacrament of communion is open to all, during their weekly food pantry no information is collected from participants. Instead, "right around the same altar where Saint Gregory's offers communion, we give away free groceries to hundreds of hungry families. We provide literally tons of fresh fruits and vegetables, rice, beans, pasta, cereal, bread; and we share our peaceful, beautiful space" (St. Gregory's Episcopal Church 2015). Volunteers at the pantry come from the population receiving food from the pantry, and volunteers prepare a communal lunch so that there is time for this mixed-group of volunteers to spend time together. The open religiosity of the food pantry that Miles recounts in her memoir speaks to clear articulation of the faith and service, but certainly could ostracize non-religious participants. Similarly, the food shelves' decision to forgo government commodity food gives them the ability to not collect income information from participants, but this in turn demands greater fundraising to replace this food source. Read in this manner, the diverse leadership teams at SHARE and RFP create a space for the development of weak theory and a philosophy of food provision grounded in the lives of both participants and committed volunteers.

While it is possible for "weak theory" to emerge from unorthodox places, this potentiality should be tempered by the idea that sites are not all unique, instead they exhibit signs of interactions with other sites, patterns formed through shared co-spatiality and existence. As such, there is a complex scalar architecture that delineates both how spaces in the voluntary sector are created and the form that co-option can take when instigated at the local level (Williams et al. 2012; Jenkins 2005). Nik Heynen draws attention to how an ideological commitment to the transnational project of black liberation allowed for the Black Panther Free Breakfast program to have currency far from the physical locations of distribution:

The ways in which the BPP struggled for social reproduction through their Breakfast Program in their Black communities and how it allowed them to organize chapters across the United States, and then produce an internationally recognized moment of revolutionary potential, exhibits how individual actors transform and re-produce the material foundations of life in scaled ways, and transform the geographies of survival.

(2009, 410)

In contrast, Johnston and Baker (2005) note that while the Good Food Box (GFB)—a Toronto-based project that provided local produce to food-insecure consumers—could survive in Toronto, it was difficult to "scale-up" or replicate this success in other areas because of the unique relationships that arose through partnerships between farmers, Toronto municipal authorities, and the non-profit sector. The inability for such a project to grow reinforces the need for a locally contingent but theoretically profound organizing principle for civil society food projects.

Be Responsive to Participants' Voices

In designing a food system for people experiencing food insecurity, it is imperative that the voices of those for who the food system is not working are heard the loudest (Pine and de Souza 2013; Adams et al. 2012). In all of the programs studied, volunteer and program managers were more food secure than program participants, and none had specific processes in place to ensure that those most affected by food insecurity could oversee and design the food program set up for their nourishment. Many community-based institutions and organizations in the "helping professions" of social work and community aid do more to create community *among volunteers* than *within* the communities in which they work (McKnight and Kretzmann 1993; also see Poppendieck 1999). Food shelves and other civil society responses to food insecurity are often places where volunteers spend meaningful time with one another while serving program participants, as opposed to places where program volunteers and participants create bonds, or places where participants are empowered to advocate for their own needs. Most food shelves could be described as spaces of anti-community for participants: people file in to receive their allotment of discounted food, and file out. In contrast, program volunteers—often organized through churches and other civic organizations—spend time with one another and are nourished spiritually through their involvement in the food programs.

Listening to the voices of people experiencing food insecurity is difficult: as a result of the often chaotic and difficult conditions of their life they do not have time for committee meetings and rarely serve as volunteers at the organizations from which they receive food. Therefore, organizers need to be proactive and search for ways to make their programs less based on the charity model. Community Action Duluth hosts Big View forums that are designed to bring people in poverty in contact with policy makers and members of the wider community in order to promote solutions to problems that can only be identified through cross-class discussion. Similarly, the Lincoln Park Fair Food Access campaign is composed of community organizers, neighborhood residents, academics, and non-profit organizations working together to find ways to bring healthier food options into Lincoln Park, a low-income neighborhood often described as a food desert (Pine and Bennett 2013).

It is especially important to bring people experiencing food insecurity into the conversation about strategies to improve food access because without their voices in the conversation their views will be misrepresented in favor of other agendas. Julie Guthman, for example, in her work on the problems inherent in "bringing good food to people," notes that while obesity might be an important issue to many in the public health and good food movement, it is not a pressing concern for people experiencing food insecurity (2008, 2013). Simply how we define the term food security is indicative of biases, thus work with those experiencing hunger often starts with presupposed understanding of the problem, which inevitably lead to conclusions that align with the interest of researchers, activists, and organizations (Bastian and Coveney 2013). To this end, in interviews with food shelf clients, many considered themselves adept cooks and did not see the same urgency for cooking classes as others in the health-promotion community. However, healthy cooking and education about the negative aspects of processed food is a common component of many food movement activities. Adams et al. (2012) in their work on PhotoVoice found that bringing clients into the conversation on program design led to innovative and unique program design choice (this finding is similar to Dutta et al. 2013). This call for participant's voices is similar to Goodman et al.'s (2011) call for "reflexivity" in the creation of alternative food projects: food access programs should involve the hungry in conversations with their clients about what "hungry" means to people experiencing food insecurity, and asking them how they define transit problems, rather than drawing on the sometimes misguided notions of programs volunteers and designers. Using the Community Based Participatory Research (CBPR) model, which demands that researchers work in close cooperation with research "subjects" to design and implement their research, Jacobs et al. (2009) created the Finding Solutions to Food Insecurity project and found that when people experiencing food insecurity create policy they pushed for a living wage, affordable housing, easier paperwork when applying for government benefits, and stronger transitional support for those moving towards self-sufficiency. This project addressed the power imbalance in service delivery by empowering those in poverty to address policy makers and reconceptualize the relationship between client and agency. Many of these proposals speak to my last point, the connections between food insecurity and poverty.

Speak to the Interconnections between Poverty and Food Insecurity

As Rebecca from SoS notes in the epigraph that opens this chapter, "[Food insecurity] is an economic issue that gets framed as a food issue, because it's easier to talk about food than it is to talk about economic justice." Because most people in the US provision themselves through shopping for groceries at conventional retail stores, people experiencing food insecurity usually find themselves in need of food aid because of their lack of ability to purchase the food they need. Food access is a question of income and any efforts that can be made to increase household income will bolster food security. Efforts to increase the amount and form of food subsidies obviously ameliorate some of the effects of food insecurity, and efforts to bolster these subsidies should be applauded. In fact, in 2012 only 62% of those in Minnesota eligible for SNAP benefits actually received those benefits (FRAC 2012). However, these benefits are directly focused on food and ignore the fact that low-income households—like all households—have complex budgets and the sophistication to manage budgets *as they see fit*. Many people alter their food budgets to support other spending goals (i.e. eating cheaper food in order to afford a new car), and federal benefit programs allow no flexibility and mandate that all food dollars be spent on food. Pushing for economic policies, such as a higher minimum wage, an expanded Earned Income Tax Credit (EITC), and an increase in the rental property tax deduction, puts dollars directly into the hands of low-income people and therefore increases their food security. Levkoe echoes this connection between poverty and food insecurity noting that food projects should adopt a "whole systems approach," and address "social justice, ecological sustainability, community health and democracy throughout all aspects of the food systems." CHUM and RFP certainly contain multiple elements that work to stabilize the lives of the poor, and these projects clearly address food insecurity as an interconnected part of poverty. Similarly, Cutts et al. (2011) note that food insecurity is closely linked to insecure housing in a way that constructs equity-based interventions in the housing market as steps towards greater food security. By "siloing" different aspects of poverty into distinct programs, the lived reality of people's lives is ignored.

Food is an everyday requirement that is imbued with culture, history, and identity (Peterson et al. 2014), and also directly connected to health and lifestyle. This makes working with food issues difficult because cuisine serves as a reservoir through which our understandings of society are expressed (Bell and Valentine 1997). Saul Alinsky, for example, in his discussion of how organizers have to authentically live the lives of the community within which they will be working, cautions them against being fake by telling this story:

> I remember a first meeting with Mexican-American leaders in a California barrio where they served me a special Mexican dinner. When we were halfway through I put down my knife and fork saying, "My God! Do you eat this stuff because you like it or because you have to? I think it's as lousy as the Jewish kosher crap I had to eat as a kid!" There was a moment of shocked silence and then everybody roared. Suddenly, barriers began to come down as they all began talking and laughing. They were so accustomed to the Anglo who would rave about the beauty of Mexican food even though they knew it was killing him, the Anglo who had memorized a few Spanish phrases with the inevitable *hasta la vista,* that it was a refreshingly honest experience to them.
>
> (1971, 70)

While we may question Alinsky's reading of the situation, he clearly uses cuisine not as a way of forming a food-based community or as an intervention into the healthy eating habits of Chicanos. Instead, food is used as an element of community cohesion in order to illustrate the ideals of honesty and integrity in pursuit of a different goal: grassroots organizing around the political empowerment of Chicanos in California. Food is used as a tool of community engagement and partnership, in not an end point.

Looking Forward

Utilizing the voluntary sector as a space from which to serve those in need of immediate food support creates a liminal space between the state and the citizen; here, the lifeworlds of volunteers and program participants come together to form new and unique spaces of care and social reproduction. As Herman and Yarwood (2014) write: "The liminal is a negotiation between 'what is' and 'what if' and is traditionally positioned as a transitional/transformative zone" (48). As such, this space has emancipatory potential because it is free from the overt strictures of government control and often guided by individuals and organizations that seek to radically transform society. However, this space has the potential to be particularly oppressive as it is used by the state to discipline those most in need of aid "from a distance" and often contains the preferences of volunteers, funders. This chapter is an exploration of that "what if" component of Herman and Yarwood's analysis of liminality: "what if we designed food access to create something new?"

The everydayness of food access projects means that they are not temporary sites of protest but instead crucial locations where the concrete acts of social reproduction take place. Their ubiquity is a sign both of care of those in civil society, and the larger forms of disrespect that marginalized people experience. As Linda from CHUM states, the repetitive nature of this work produces a sense of dismay:

> Well, you know why are we still doing this? Why haven't we figured this out? I mean, this has been over 30 years this place has been opened and what have we done? We just keep serving more people. And I am not saying we at CHUM are wrong, [but] what are we as people doing that we are okay with that being there, you know?

Ten, 20, 30 years from now, I fear that committed food activists will be asking themselves the same question. As opposed to using food access as a way of reiterating "what is," it is incumbent to provide food while assertively asking "what if?"

Note

1   For an interesting food-regime-based analysis of "nutritionalisation" see Dixon, 2009.

Bibliography

Abraham, John. 1991. *Food and Development*. London: World Wide Fund for Nature and Kogan Page Ltd.
Adams, Karen, Cate Burns, Anna Liebzeit, Jodie Ryschka, Sharon Thorpe, and Jennifer Browne. 2012. "Use of Participatory Research and Photo-voice to Support Urban Aboriginal Healthy Eating." *Health and Social Care in the Community* 20(5): 497–505.
Agamben, Giorgio 1993. *The Coming Community*. Translated by Michael Hardt. Minneapolis: University of Minnesota Press.
Agyeman, Julian and Jesse McEntee. 2014. "Moving the Field of Food Justice Forward Through the Lens of Urban Political Ecology." *Geography Compass* 8(3): 211–220.
Alexander, Michelle. 2012. *The New Jim Crow: Mass Incarceration in the Age of Colorblindness*. New York: The New Press.
Alinsky, Saul. 1971. *Rules for Radicals: A Pragmatic Primer for Realistic Radicals*. New York: Random House.
Alkon, Alison Hope and Christie Grace McCullen. 2011. "Whiteness and Farmers Markets: Performances, Perpetuation…Contestations?" *Antipode* 43(4): 937–959.
Alkon, Alison Hope and Teresa Marie Mares. 2012. "Food Sovereignty in US Food Movements: Radical Visions and Neoliberal Constraints." *Agriculture and Human Values* 29: 347–359.
Alkon, Alison Hope. 2012. *Black, White, and Green.* Athens: University of Georgia Press.
Allen, Patricia and Julie Guthman. 2006. "From 'Old School' to 'Farm-to-school' Neoliberalization from the Ground Up." *Agriculture and Human Values* 23(4): 401–415.
Allen, Patricia. 1999. "Reweaving the Food Security Safety Net: Mediating Entitlement and Entrepreneurship." *Agriculture and Human Values* 16: 117–129.
Allen, Patricia. 2004. *Together at the Table*. State College: Pennsylvania State University.
Allen, Will. 2013. *The Good Food Revolution: Growing Healthy Food, People, and Communities*. New York: Gotham.
Anater, Andrea, Rita McWilliams, and Varl Latkin. 2011. "Food Acquisition Practices Used by Food-insecure Individuals When They Are Concerned About Having Sufficient Food for Themselves and Their Households." *Journal of Hunger and Environmental Nutrition* 6(1): 27–44.
Anderson, Benedict. 1983. *Imagined Communities: Reflections on the Origin and Spread of Nationalism*. London: Verso.

Atia, Mona. 2012. "'A Way to Paradise': Pious Neoliberalism, Islam, and Faith-based Development." *Annals of the Association of the American Geographers* 102(4): 808–827.
Atkins, Peter and Ian Bowler. 2001. *Food in Society: Economy, Culture, Geography*. London: Arnold Publishers.
Backett-Milburn, Kathryn, Wendy Wills, Mei-Lei Roberts, and Julia Lawton. 2010. "Food and Family Practices: Teenagers, Eating, and Domestic Life in Differing Socio-economic Circumstances." *Children's Geographies* 8(3): 303–314.
Ballamingie, Patricia and Sarah Walker. 2013. "Field of Dreams. Just Food's Proposal to Build a Community Food and Sustainable Agriculture Hub in Ottawa, Ontario." *Local Environments* 18(5): 529–542.
Bartlett, Peggy. 2013. "Campus Sustainable Food Projects: Critique and Engagement." *American Anthropologist* 113(1): 101–115.
Bastian, Amber and John Coveney. 2013. "The Responsibilisation of Food Security: What is the Problem Represented to be?" *Health Sociology Review* 22(2): 162–173.
Bedford-Strohm, Heinrich. 2012. "Food Justice and Christian Ethics." *Verbum et Ecclesia* 33(2):1–6.
Bee, Beth. 2011. "Gender, Solidarity, and the Paradox of Microfinance: Reflections from Bolivia." *Gender, Place & Culture* 18(1): 23–43.
Bell, Colin and Howard Newby. 1978. *Community Studies*. London: George Allen and Unwin.
Bell, David and Gill Valentine. 1997. *Consuming Geographies: We Are Where We Eat*. Oxford: Routledge.
Berner, Maureen, Trina Ozer and Sharon Paynter. 2008. "A Portrait of Hunger, the Social Safety Net, and the Working Poor." *The Policy Studies Journal* 36(3).
Black, Daniel and Joanne Kouba. 2005. "A Comparison of the Availability and Affordability of a Market Basket in Two Communities in the Chicago Area." *Public Health Nutrition* 9(7): 837–845.
Blanchard, Troy and Thomas Lyson. 2002. "Access to Low-cost Groceries in Non-metropolitan Counties: Large Retailers and the Creation of Food Deserts." Paper presented at Measuring Rural Diversity Conference, Washington DC, November 21–22.
Boal, Augusto. 1979. *Theater of the Oppressed*. New York: Theater Communication Group.
Boero, Natalie. 2010. "Fat Kids, Working Moms, and the Epidemic of Obesity Race, Class and Mother-blame." In *The Fat Studies Reader* edited by Esther Rothblum, Sondra Solovay, and Marilyn Wann. 113–119. New York: NYU Press.
Bondi, Liz. 2005. "Working the Spaces of Neoliberal Subjectivity: Psychotherapeutic Technologies, Professionalisation, and Counseling." *Antipode* 37(3): 497–514.
Bourdieu, Pierre. 1993. *The Field of Cultural Production*. Cambridge: Polity.
Brenner, Neil and Nik Theodore. 2002. "Cities and the Geographies of 'Actually Existing Neoliberalism'" in *Spaces of Neoliberalism: Urban Restructuring in North America and Western Europe* edited by Neil Brenner and Nik Theodore, 1–32. Chichester, UK: John Wiley & Sons, Ltd.
Brown, Michael. 1997. *RePlacing Citizenship: AIDS Activism and Radical Democracy*. New York: Guilford Press.
Burchell, Graham, Colin Gordon, and Peter Miller. 1991. *The Foucault Effect: Studies in Governmentality*. Chicago: University of Chicago Press.

Buttle, Martin. 2008. "Diverse Economies and the Negotiations and Practices of Ethical Finance: The Case of Charity Bank." *Environment & Planning A* 40(9): 2097–2113.

Carney, Megan. 2012. "Compounding Crises of Economic Recession and Food Insecurity: A Comparative Study of Three Low-income Communities in Santa Barbara County." *Agriculture and Human Values* 29: 185–201.

Carolan, Michael. 2011. *Embodied Food Politics*. Burlington, Vermont: Ashgate.

Carolan, Michael. 2013. *The Real Cost of Cheap Food*. New York: Earthscan.

Chatterton, Paul. 2005. "Making Autonomous Geographies: Argentina's Popular Uprising and the 'Movimiento de Trabajadores Desocupados' (Unemployed Workers Movement)." *Geoforum* 36: 545–561.

Chatterton, Paul and Jenny Pickerill. 2010. "Everyday Activism and Transitions Towards Post-capitalist Worlds." *Transactions of the Institute of British Geographers* 475–505.

Chilton, Mariana and Donald Rose. 2009. "A Rights-based Approach to Food Insecurity in the United States." *American Journal of Public Health* 99(7): 1203–1211.

CHUM. 2013. Annual Report. Duluth, Minnesota.

ChurchMouse Chronicles. 2014. "Welcome to the ChurchMouse Chronicles." Accessed online on February 2nd, 2015. http://www.churchmouse.net/images/english.htm.

Cloke, Paul, Jon May, and Sarah Johnsen. 2011. *Swept up Lives? Re-envisioning the Homeless City*. New Jersey; Wiley.

Coleman-Jensen, Christian Gregory, and Anita Singh. Household Food Security in the United States in 2013. ERS Report Number 173. USDA / Economic Research Service Washington DC. Bulletin 173.

Coleman-Jensen, Alisha, William McFall, and Mark Nord. 2013. "Food Insecurity in Households With Children: Prevalence, Severity, and Household Characteristics, 2010–11." USDA / Economic Research Service Washington DC. Bulletin 113.

Congressional Budget Office. 2010. *The Distribution of Household Income and Federal Taxes, 2010*. Washington DC.

Conlan, Timothy. 1998. *From New Federalism to Devolution: Twenty-five Years of Intergovernmental Reform*. Washington DC: The Brookings Institution.

Conradson, David. 2003. "Doing Organisational Space: Practices of Voluntary Welfare in the City." *Environment and Planning A* 35: 1975–1992.

Cox, Rosie, Lewis Holloway, Laura Venn, Liz Dowler, Jane Ricketts Hein, Moya Kneafsey, and Helena Tuomainen. 2008. "Common Ground? Motivations for Participation in a Community Supported Agriculture Scheme." *Local Environment* 13(3):203–218.

Cramer, Janet, Carlnita Greene, and Lynne Walters. 2011. *Food as Communication: Communication as Food*. Berne, Switzerland: Peter Lang Academic Publishers.

Cutts, Diana Becker, Alan Meyers, Maureen Black, Patrick Casey, Mariano Chilton, John Cook, Joni Geppert, Stephanie Ettinger de Cuba, Timothy Heeren, Sharon Coleman, Ruth Rose-Jacobs, and Deborah Frank. 2011. "US Housing Insecurity and the Health of Very Young Children." *American Journal of Public Health*. 101(8): 1508–1514.

D'Sylva, Andrea and Brenda Beagan. 2011. "'Food is Culture, but it's Also Power': The Role of Food in Ethnic and Gender Identity Construction Among Goan Canadian Women." *Journal of Gender Studies* 20(3): 279–289.

Davis, Richard and James Collins. 2014. "Layers of Inequality: Power, Policy, and Health." *American Journal of Public Health* 104(S1): 8–10.

Day, Dorothy. 1963. *Loaves and Fishes*. New York: Harper and Row.

Daya, Shari and Raksha Authar. 2012. "Self, Others and Objects in an 'Alternative Economy': Personal Narratives from the Heiveld Rooibos Cooperative." *Geoforum* 43(5): 885–893.

Dean, Mitchell. 1999. *Governmentality: Power and Rule in Modern Society.* London: Sage.
Debord, Guy. 1983[1967]. *Society of the Spectacle.* Detroit, Michigan: Black and Red.
Derickson, Kate Driscoll. 2009. "Towards a Non-totalizing Critique of Capitalism." *The Geographical Bulletin* 50: 3–15.
De Schutter, Olivier. 2012. "From Charity to Entitlement: Implementing the Right to Food in Southern and Eastern Africa." Briefing Note 05. UN: Geneva, Switzerland.
DeVerteuil, Geoffrey, Woobae Lee, and Jennifer Wolch. 2002. "New Spaces for the Local Welfare State?" The Case of General Relief in Los Angeles County." *Social and Cultural Geography* 3.3:229–246.
Dixon, Jane. 2009. "From Imperial to the Empty Calorie: How Nutrition Relations Underpin Food Regime Transitions" *Agriculture and Human Values* 26: 321–333.
Dixon, Jane. 2011. "Diverse Food Economies, Multivariant Capitalism, and the Community Dynamic Shaping Contemporary Food Systems." *Community Development Journal* 46: 20–35.
Donald, Betsy and Alison Blay-Palmer. 2006. "The Urban Creative-food Economy: Producing Food for the Urban Elite or Social Inclusion Opportunity." *Environment and Planning A* 38: 1901–1920.
Doonan, Natalie. 2014. "A Study in Dissonance: Performing Alternative Food Systems." *Canadian Theatre Review* 157: 39–42.
Dreier, Peter, John Mollenkopf, and Todd Swanstrom. 2004. *Place Matters: Metropolitics for the Twenty-first Century.* Lawrence, Kansas: University of Kansas.
Duda, Mark, Martin Jones, and Andrea Criscione. 2010. *The Sportsman's Voice: Hunting and Fishing in America.* State College, Pennsylvania: Venture Publishing.
Dutta, Mohan Jyoti, Agaptus Anaele, and Christina Jones. 2012. "Voices of Hunger: Addressing Health Disparities Through the Culture-centered Approach." *Journal of Communication* 63(1): 159–180.
Dwyer, Owen and John Paul Jones III. 2000. "White Socio-spatial Epistemology. *Social and Cultural Geography* 1(2): 209–222.
Dyer, Jeffrey, Hal Gregersen, and Clayton Christensen. 2008. "Entrepreneur Behaviors, Opportunity Recognition, and the Origins of Innovative Ventures." *Strategic Entrepreneurship Journal* 2: 317–38.
Eaton, Randall. 2008. "Modern Hunters are Stewards of Wildlife and the Environment." In *Hunting* Edited by Dawn Laney, 38–46. Farmington Hills: The Gale Group.
Ehrlich, Paul. 1968. *The Population Bomb.* New York: Ballantine Books.
Elwood, Sarah. 2006. "Beyond Cooptation or Resistance: Urban Spatial Politics, Community Organizations, and GIS-based Spatial Narratives." *Annals of the Association of American Geographers* 96(2): 323–341.
Endale, Worku, Zelalem Birhanu Mengesha, Azeb Atinafu, and Akilew Awoke Adane. 2014. "Food Insecurity in Farta District, Northwest Ethiopia: A Community-based cross-sectional Study. *BMC Research Notes* 7: 130–136.
Environmental Working Group. 2013. "Farm Subsidies: The United States Summary Information." Accessed online February 6th, 2015. http://farm.ewg.org/region.php?fips=00000&statename=theUnitedStates.
Escobar, Arturo. 1995. *Encountering Development: The Making and Unmaking of the Third World.* Princeton: Princeton University Press.
Etzioni, Amitai. 2004. "On Virtual, Democratic Communities." In *Community in the Digital Age.* Edited by Andrew Feenberg and Darin Barney. 225–238. London: Rowan and Littlefield.

Evers, Anna and Nicole Hodgson. 2011. "Food Choice and Local Food Access Among Perth's Community Gardeners." *Local Environment* 16(6): 585–602.

Farmer, Paul. 2003. *Pathologies of Power: Health, Human Rights, and the New War on the Poor*. Berkeley: University of California Press.

Feeding American. 2013. 2013 Annual Report: Solving Hunger Together. Washington DC: Feeding America.

Fielding-Miller, Rebecca, Zandile Mnisi, Darrin Adams, Stefan Baral, and Caitlin Kennedy. 2014. "'There is Hunger in My Community': A Qualitative Study of Food Security as a Cyclical Force in Sex Work in Swaziland." *BMC Public Health*. 14:79–88.

Firth, Jeanne. 2012. "Healthy Choices and Heavy Burdens: Race, Citizenship, and Gender in the 'Obesity Epidemic.'" *Journal of International Women's Studies* 13(2): 33–50.

Flynn, Mary, Steven Reinert, and Andrew Schiff. 2013. "A Six-week Cooking Program of Plant-based Recipes Improves Food Security, Body Weight, and Food Purchases for Food Pantry Clients." *Journal of Hunger and Environmental Nutrition* 8(1): 73–84.

Ford, James, Marie-Pierre Lardeau, Hilary Blackett, Susan Chatwood, and Denise Kurszewski. 2013. "Community Food Programs Use in Inuvik Northwest Territories." *BMC Public Health* 13: 970–985.

Fothergill, Alice. 2003. "The Stigma of Charity: Gender, Class, and Disaster Assistance." *The Sociological Quarterly* 44(4):659–680.

Foucault, Michel. 1975. *Discipline and Punish*. New York: Vintage Books.

Frankenberg, Ruth. 1993. *The Social Construction of Whiteness: White Women, Race Matters*. London: Routledge.

Franklin, Brandi, Ashley Jones, Dejuan Love, Stephanie Puckett, Justin Macklin, and Shelley White-Means. 2012. "Exploring Mediators of Food Insecurity and Obesity: A Review of Recent Literature." *Journal of Community Health* 37: 253–264.

Freire, Paulo. 1970. *Pedagogy of the Oppressed*. New York: Continuum.

Friedmann, Harriet. 2007. "Scaling Up: Bringing Public Institutions and Food Service Corporations into the Project for a Local, Sustainable Food System in Ontario. *Agriculture and Human Values* 24: 389–398.

Friedmann, John. 1992. *Empowerment: The Politics of Alternative Development*. Cambridge, Massachusetts: Blackwell.

Fukuyama, Francis 1992. *The End of History and the Last Man*. New York: Free Press.

Fyfe, Nicholas and Christine Milligan. 2003 "Out of the Shadows: Exploring Contemporary Geographies of Voluntarism." *Progress in Human Geography* 27(4): 397–413.

Fyfe, Nicholas. 2005. "Making Space for 'Neo-communitarianism'? The Third Sector, State and Civil Society in the UK." *Antipode* 37(3): 536–557.

Galt, Ryan. 2013. "The Moral Economy is a Double-edged Sword: Explaining Farmers' Earnings and Self-exploitation in Community Supported Agriculture." *Economic Geography* 89(4): 341–365.

Ghose, Rina and Margaret Pettygrove. 2014. "Urban Community Gardens as Spaces of Citizenship." *Antipode* 46(4):1092–1112.

Gibson-Graham, J.K., 2006. *A Postcapitalist Politics*. Minneapolis: University of Minnesota Press.

Gibson-Graham, J.K., Jenny Cameron, and Stephen Healy. 2013. *Take Back the Economy: An Ethical Guide for Transforming our Communities*. Minneapolis: University of Minnesota Press.

Gibson-Graham, J.K. 1996. *The End of Capitalism (As We Knew It): A Feminist Critique of Political Economy*. Minneapolis: University of Minnesota Press.

Gilmore, Ruth Wilson. 2007. *Golden Gulag: Prisons, Surplus, Crisis, and Opposition in Globalizing California.* Los Angeles: University of California Press.

Goodman, Michael, Damian Maye, and Lewis Holloway. 2010. "Ethical Foodscapes? Premises, Promises, and Possibilities." *Environment and Planning A* 42(8): 1782–96.

Goodman, David, E. Melanie DuPuis, and Michael K. Goodman. 2012. *Alternative Food Networks: Knowledge, Practice, and Politics.* London: Routledge.

Gottlieb, Robert and Anupama Joshi. 2010. *Food Justice: Food, Health, and the Environment.* Boston: MIT Press.

Gray, Mel. 2010. "Social Development and the Status Quo: Professionalisation and Third Way Co-optation." *International Journal of Social Welfare* 19: 463–470.

Griffin, Michael and Jennie Weiss Block. 2013. "Introduction." In *In the Company of the Poor: Conversations with Dr. Paul Farmer and Fr. Gustavo Gutiérrez.* Edited by Michael Griffin and Jennie Weiss Block, 1–14. Maryknoll, New York: Orbis Books.

Grigg, David. 1985. *The World Food Problem.* Oxford: Basil Blackwell.

Gross, Joan. 2009. "Capitalism and Its Discontents: Back-to-the-Lander and Freegan Foodways in Rural Oregon." *Food & Foodways: History & Culture of Human Nourishment* 17(2): 57–79.

Guha, Ramachandra. 2006. *How Much Should One Person Consume: Environmentalism in India and the US.* Berkeley: University of California Press.

Guthman, Julie. 2003. "Fast Food/Organic Food: Reflexive Taste and the Making of 'Yuppie Chow.'" *Social and Cultural Geography* 4(1): 45–58.

Guthman, Julie. 2008. "Bringing Good Food to Others: Investigating the Subjects of Alternative Food Practice." *Cultural Geographies* 15(4): 431–447.

Guthman, Julie. 2011. *Weighing In: Obesity, Food Justice, and the Limits of Capitalism.* Los Angeles: University of California Press.

Gutiérrez, Gustavo. 2003. *We Drink from Our Own Wells: The Spiritual Journey of a People.* Maryknoll, NY: Orbis Books.

Hackworth, Jason and Joshua Akers. 2010. "Faith in the Neoliberlisation of Post-Katrina New Orleans." *Tijdschrift Voor Economische enSociale Geografie* 10(1): 39–54.

Hall, Timothy. 2006. *Changing the Face of Hunger.* Nashville, Tennessee: Thomas Nelson.

Haney, Lynne. "Gender, Welfare, and State Punishment." *Social Politics: International Studies in Gender, State, and Society* 11(3): 333–362.

Harder, Miriam and George Wenzel. 2012. "Inuit Subsistence, Social Economy, and Food Security in Clyde River, Nunavut." *Arctic* 65(3): 305–318.

Harris, Edmund. 2009. "Neoliberal Subjectivities or a Politics of the Possible? Reading for Difference in Alternative Food Networks." *Area* 55–63.

Harvey, David. 1974. "The Political Implications of Population-resources Theory." *Economic Geography* 50(3): 256–77.

Harvey, David. 1991. *The Condition of Postmodernity: An Enquiry into the Origins of Cultural Change.* New Jersey: Wiley Blackwell.

Harvey, David. 2000. *Spaces of Hope.* Los Angeles: University of California Press.

Harvey, David. 2007. *A Brief History of Neoliberalism.* Oxford: Oxford University Press.

Hawkes, Corinna and Jacqui Webster. 2000. *Too Much* and *Too Little: Debates on Surplus Food Redistribution.* London: Sustain: The Alliance for Better Food and Farming.

Hayes-Conroy, Alison and Jessica Hayes-Conroy. 2010. "Visceral Difference: Variations in Feeling (Slow) Food." *Environment and Planning A* 42: 2956–2971.

Hébert, Karen and Diana Mincyte. 2014. "Self-Reliance Beyond Neoliberalism: Rethinking Autonomy at the Edges of Empire." *Environment and Planning D: Society and Space* 32: 206–222.

Heldke, Lisa. 2009. "Three Social Paradigms for Access: Charity, Rights, and Coresponsibility." In *Critical Food Issues: Problems and State-of-the-Art Solutions*, Edited by Lynn Walter and Laurel Phoenix. Santa Barbara: Praeger.

Henderson, George. 2004. "'Free' Food, the Local Production of Worth, and the Circuit of Decommodification: A Value Theory of the Surplus." *Environment and Planning D* 22: 485–512.

Hendrickson, Deja, Chery Smith, and Nicole Eikenberry. 2006. "Fruit and Vegetable Access in Four Low-income Food Desert Communities in Minnesota." *Agriculture and Human Values* 23: 371–83.

Herman, Agatha and Richard Yarwood. 2014. "From Warfare to Welfare: Veterans, Military Charities, and the Blurred Spatiality of Post-service Welfare in the United Kingdom." *Environment and Planning A* 47: 2628–2644.

Heynen, Nik. 2009. "Bending the Bars of Empire from Every Ghetto to Feed the Kids: The Black Panther Party's Radical Anti-hunger Politics of Social Reproduction and Scale." *The Annals of the Association of American Geographers.* 99(2): 406–422.

Heynen, Nik. 2010. "Cooking up Non-violent Civil Disobedient Direct Action for the Hungry: Food Not Bombs and the Resurgence of Radical Democracy." *Urban Studies* 47(6): 1225–1240.

Hinrichs, C. Clare. 2000. "Embeddedness and Local Food Systems: Notes on Two Types of Direct Agricultural Market." *Journal of Rural Studies* 16: 295–303.

Hoisington, Anne, Jill Schultz, and Sue Butkus. 2002. "Coping Strategies and Nutritional Education Needs Among Food Pantry Users." *Journal of Nutrition Education and Behavior* 34: 326–333.

Holloway, Lewis, Moya Kneafsey, Laura Venn, Rosie Cox, Elizabeth Dowler, and Helena Tuomainen. 2007. "Possible Food Economies? A Methodological Framework for Exploring Food Production-consumption Relationships." *European Journal for Rural Sociology* 47(1): 1–19.

Holloway, Sarah and Helena Pimlott-Wilson. 2012. "Neoliberalism, Policy Localisation, and Idealised Subjects: A Case Study on Education Restructuring in England." *Transactions of the Institute of British Geographers* 37: 639–654.

Holt-Giménez, Eric and Yi Wang. 2011. "Reform or Transformation? The Pivotal Role of Food Justice in the U.S. Food Movement." *Race/Ethnicity: Multidisciplinary Global Contexts* 5(1): 83–102.

Home and Away Ministries. 2009. "Home and Away Ministries." Accessed online on February 6th, 2015. http://www.homeandawayministries.org/.

Hooks, Bell. 1994. *Teaching to Transgress: Education as the Practice of Freedom*. New York: Routledge.

Horwitt, Sanford. 1992. *Let Them Call Me Rebel: Saul Alinsky His Life and Legacy*. New York: Vintage Books.

Hudelson, Richard and Carl Ross. 2006. *By the Ore Docks: A Working People's History of Duluth*. Minneapolis: University of Minnesota Press.

Huffman, Sonya Kostova and Helen Jensen. 2003. Do Food Assistance Programs Improve Household Food Security? Recent Evidence from the United States. Working Paper 03-WP 335, Center for Agricultural and Rural Development, Iowa State University.

Institute for Sustainable Futures. 2013. The Western Lake Superior Good Food Movement: 2013 Status Report. Duluth: Institute for Sustainable Futures.

Jacobs, Maxine, Kate Pruitt-Chapin, and Chris Rugeley. 2009. "Towards Reconstructing Poverty Knowledge: Addressing Food Insecurity Through Grassroots Research Design and Implementation." *Journal of Poverty* 3: 1–19.

Jarosz, Lucy. 2008 "The City in the Country: Growing Alternative Food in Networks in Metropolitan Areas." *Journal of Rural Studies* 24: 231–244.

Jenkins, Katy. 2005. "No Way Out? Incorporating and Restructuring the Voluntary Sector Within Spaces of Neoliberalism." *Antipode* 37(3): 613–618.

Jessop, Bob and Ngai-Ling Sum. 2000. "An Entrepreneurial City in Action: Hong Kong's Emerging Strategies in and for (inter)urban Competition." *Urban Studies* 37(12): 2287-2313.

Johnston, Josée and Lauren Baker. 2005. "Eating Outside the Box: FoodShare's Good Food Box and the Challenge of Scale." *Agriculture and Human Values* 22: 313–325.

Judd, Dennis and Susan Fainstein. 1999. *The Tourist City*. New Haven: Yale University Press.

Kaplan, Robert. 1994. "The Coming Anarchy: How Scarcity, Crime, Overpopulation, Tribalism, and Disease are Rapidly Destroying the Social Fabric of Our Planet." *The Atlantic* February 1.

Karger, Howard and David Stoesz. 2009. *American Social Welfare Policy*. New York: Routledge.

Katz, Cindi. 2001. "Vagabond Capitalism and the Necessity of Social Reproduction." *Antipode* 33: 709–728.

Katz, Cindi. 2004. *Growing Up Global: Economic Restructuring and Children's Everyday Lives*. Minneapolis: University of Minnesota Press.

Keil, Roger. 2002. "'Common-sense' Neoliberalism: Progressive Conservative Urbanism in Toronto, Canada." *Antipode* 34(3): 578–601.

Kendall, Jeremy and Martin Knapp. 1995. "Measuring the Performance of Voluntary Organizations." *Public Management* 2(1): 105–132.

Kingsolver, Barbara. 2007. *Animal, Vegetable, Miracle*. New York: HarperCollins.

Kobayashi, Audrey and Linda Peake. 2000. "Racism Out of Place: Thoughts on Whiteness and an Anti-racist Geography in the New Millennium." *Annals of the Association of American Geographers* 90(2): 392–403.

Kropotkin, Peter. 1906. *The Conquest of Bread*. New York: G. P. Putnam's Sons.

Kropotkin, Peter. 1955[1902]. *Mutual Aid: A Factor of Evolution*. Boston: Extending Horizons Press.

Krugman, Paul. 2007. *The Conscience of a Liberal*. New York: W. W. Norton & Company.

Lappé, Frances Moore and Joseph Collins. 1977. *Food First*. Boston: Houghton Mifflin Company.

Larner, Wendy. 2000. "Neo-liberalism: Policy, Ideology, Governmentality." *Studies in Political Economy* 63: 5–25.

Larsen, Kristian, and Jason Gilliland. 2008. "Mapping the Evolution of 'Food Deserts' in a Canadian City: Supermarket Accessibility in London, Ontario, 1961–2005." *International Journal of Health Geographics* 7(16): 1–16.

Laverack, Glen. 2004. *Health Promotion Practice: Power and Empowerment*. London: Sage.

Lederman, Jacob. 2014. "Urban Fads and Consensual Fictions: Creative, Sustainable, and Competitive Policies in Buenos Aires." City & Community 14(1): 47-67.

Lepofsky, James and James Fraser. 2003, "Building Community Citizens: Claiming the Right to Place-making in the City'." *Urban Studies* 40(1): 127–142.

Levkoe, Charles. 2011. "Towards a Transformative Food Politics." *Local Environment* 16(7): 687–705.

Lewis, Nick, Owen Lewis, and Yvonne Underhill-Sem. 2009. "Filling Hollowed-out Spaces with Localised Meanings, Practices, and Hope: Progressive Neoliberal Spaces in Te Rarawa." *Asia Pacific Viewpoint* 50(2): 166–184.

Link, Bruce and Jo Phelan. 2001. "Conceptualizing Stigma." *Annual Review of Sociology* 27: 363–385.

Lipsky, Michael. 1980. *Street-level Bureaucracy*. New York: Russell Sage Foundation.

Little, Ruth, Damian Maye, and Brian Ilbery. 2010. "Collective Purchase: Moving Local and Organic Foods Beyond the Niche Market." *Environment & Planning A* 42(8): 1797–1813.

Lutz, Catherine. 2002. *Homefront: A Military City and the American Twentieth Century*. Boston: Beacon Press.

Macleavy, Julie. 2007. "The Six Dimensions of New Labour: Structures, Strategies, and Languages of Neoliberal Legitimacy." *Environment and Planning A* 39: 1715–1734.

Macleavy, Julie and Columba Peoples. 2009. "Workfare-welfare: Neoliberalism, 'Active' Welfare, and the New American Way of War." *Antipode* 41(5): 890–915.

Maron, Asa. 2012. "Conflicting Articulations of Citizenship Under a Neoliberal State Project: The Contested Implementation of the Israeli Workfare Programme." *Mediterranean Politics* 17(3): 427–445.

Maurin, Peter. 1949. *Catholic Radicalism: Phrased Essays for the Green Revolution*. New York: Catholic Worker Books.

McCann, Eugene. 2011. "Urban Policy Mobilities and Global Circuits of Knowledge: Toward a Research Agenda." *Annals of the Association of American Geographers* 101(1): 107-130.

McKnight, John L. and John P. Kretzmann. 1993. *Building Communities from the Inside Out: A Path Toward Finding and Mobilizing a Community's Assets*. Illinois: ACTA Press.

McKnight, John. 1995. *The Careless Society: Community and its Counterfeits*. New York: Basic Books.

McMichael, Phillip. 2009. "A Food Regime Analysis of the World Food Crisis." *Agriculture and Human Values* 4: 281–295.

Mercer, Kobena. 1992. "'1968': Periodizing Postmodern Politics and Identity." In *Cultural Studies,* edited by Lawrence Grossberg, Cary Nelson, and Paula A. Treichler, pages 424–449. New York: Routledge.

Merrifield, Andy. 2013. *The Politics of the Encounter: Urban Theory and Protest Under Planetary Urbanization*. Athens: University of Georgia Press.

Michimi, Akihiko and Michael Wimberly. 2010. "Associations of Supermarket Accessibility With Obesity and Fruit and Vegetable Consumption in the Conterminous United States." *International Journal of Health Geographics* 9(49): 1–14.

Miewald, Christiana and Eugene McCann. 2014. "Foodscapes and the Geographies of Poverty: Sustenance, Strategy, and Politics in an Urban Neighborhood." *Antipode* 46(2): 537–556.

Miles, Sara. 2007. *Take This Bread: The Spiritual Memoir of a Twenty-first Century Christian*. New York: Ballantine Books.

Miller, Sally. 2003. "One Belly, One Vote: Food Democracy Lives at the Co-op." *Alternatives Journal* 29(4):18.

Morton, Lois Wright, Ella Annette Bitto, Mary Jane Oakland, and Mary Sand. 2005. "Solving the Problems of Iowa Food Deserts: Food Insecurity and Civic Structure." *Rural Sociology* 70(1): 94–112.

Musick, Marc and John Wilson. 2008. *Volunteers: A Social Profile*. Indianapolis: Indiana University Press.
Newman, Kathe and Robert W. Lake. 2006. "Democracy, Bureaucracy, and Difference in Community Development Politics Since 1968." *Progress in Human Geography* 30(1): 1–18.
Nord, Mark and Anne Marie Golla. 2009. "Does SNAP Decrease Food Insecurity? Untangling the Self-selection Effect." USDA Economic Research Service: Washington DC.
Ong, Aihwa. "Cultural Citizenship as Subject-making: Immigrants Negotiate Racial and Cultural Boundaries in the US." *Current Anthropology* 37(5): 737–762.
Orfield, Myron. 2002. *American Metropolitics: The New Suburban Reality*. Washington DC: Brookings Institute.
Orleck, Annelise and Lisa Gayle Hazirjian. 2011. *The War on Poverty: A New Grassroots History, 1964–1980."* Athens: University of Georgia Press.
Patel, Raj. 2008. *Stuffed and Starved: The Hidden Battle for the World Food System*. Brooklyn: Melville House Publishing.
Peck, Jamie and Adam Tickell. 2002. "Neoliberalizing Space." *Antipode* 34(3): 380–404.
Peck, Jamie. 2001. *Workfare States*. New York: The Guilford Press.
Peterson, Deborah and Alan Lupton. 1996. *The New Public Health: Health and Self in the Age of Risk*. London: Sage.
Peterson, Tina and Katherine Turner. 2014. "'Extravagance and Folly' Versus 'Proper Food': Domestic Scientists, Celebrity Chefs, and the Ongoing Food Reform Movement." *Journal of Popular Culture* 47(4): 817–837.
Peterson, Wesley. 2009. *A Billion Dollars a Day: The Economics and Politics of Agricultural Subsidies*. Malden, Massachusetts: Wiley-Blackwell.
Pine, Adam and Rebecca de Souza. 2014. "Including the Voices of Communities in Food Insecurity Research: An Empowerment-based Agenda for Food Scholarship." *Journal of Agriculture, Food Systems, and Community Development* ttp://dx.doi.org/10.5304/jafscd.2013.034.007: 71–79.
Pine, Adam and John Bennett. 2013. "Food Access and Food Deserts: the Diverse Methods that Residents of a Neighborhood in Duluth, Minnesota Use to Provision Themselves." *Community Development* 45(4): 317–336.
Piven, Frances Fox and Richard A. Cloward. 1971. *Regulating the Poor: The Functions of Public Welfare*. New York: Vintage Books.
Pollack, Shoshana. 2009. "Creating Submissive Subjects: Lone Mothers and Social Assistance Regimes in Canada." *Benefits* 17(3): 225–235.
Pollan, Michael. 2006. *The Omnivore's Dilemma: A Natural History of Four Meals*. New York: Penguin Press.
Poppendieck, Janet. 1999. *Sweet Charity? Emergency Food and the End of Entitlement*. New York: Penguin Press.
Poppendieck, Janet. 2014. *Breadlines Knee-deep in Wheat: Food Assistance in the Great Depression*. Berkeley: University of California Press.
Pudup, Mary Beth. 2008. "It Takes a Garden: Cultivating Citizen-subjects in Organized Garden Projects." *Geoforum* 39: 1228–1240.
Putnam, Robert. 2000. *Bowling Alone: The Collapse and Revival of American Community*. New York: Simon & Schuster.
Raco, Mike. 2005. "Sustainable Development, Rolled-out Neoliberalism, and Sustainable Communities." *Antipode* 37(2): 324–347.

Rankin, Katherine. 2001. "Governing Development: Neoliberalism, Microcredit, and Rational Economic Woman." *Economy and Society* 30(1): 18–37.

Reich, Robert B. 2012. *Beyond Outrage: What has Gone Wrong with Our Economy and Our Democracy, and How to Fix It.* New York: Vintage.

Ridzi, Frank. 2009. *Selling Welfare Reform: Work-first and the new Common Sense of Employment.* New York: NYU Press.

Robbins, Paul, Marla Emery, and Jennifer Rice. 2008. "Gathering in Thoreau's Backyard: Non-timber Forest Product Harvesting as Practice." *Area* 40(2): 265–277.

Roediger, David. 1991. *The Wages Of Whiteness.* London: Verso.

Rollinson, Paul. 1998. "The Everyday Geography of the Homeless in Kansas City." *Geografiska Annaler* 80B: 101–115.

Rose, Nikolas. 1999. *Powers of Freedom.* Cambridge: Cambridge University Press.

Rose, Nikolas. 2000. "Government and Control." *British Journal of Criminology* 40: 321–339.

Rosol, Marit. 2012. "Community Volunteering as Neoliberal Strategy? Green Space Production in Berlin." *Antipode* 44(1): 239–257.

Rubin, Herbert and Irene Rubin. 2007. *Community Organizing and Development* 4th Edition. New York: Pearson.

Santos, Boaventura de Sousa. 2007. "The World Social Forum: Toward a Counter-hegemonic Globalization." In *The World Social Forum: Challenging Empires,* edited by Jai Sen, Anita Anand, Arturo Escobar, and Peter Waterman, 235–245. Montreal: Black Rose Books.

Schulz, Amy, Graciela Mentz, Laurie Lachance, Jonetta Johnson, Causandra Gaines, and Barbara Israel. 2012. "Associations Between Socioeconomic Status and Allostatic Load: Effects of Neighborhood Poverty and Tests of Mediating Pathways. *American Journal of Public Health* 102(9): 1706–1714.

Sen, Amartya. 1981. *Poverty and Famines: An Essay on Entitlement and Deprivation.* Oxford: Clarendon Press.

Shannon, Jerry. 2014. "Food Deserts: Governing Obesity in the Neoliberal City." *Progress in Human Geography* 38(2): 248–266.

Shaw, Hilary. 2006. "Food Deserts: Towards the Development of a Classification." *Geografiska Annaler: Series B, Human Geography* 88(2): 231–247.

Shiva, Vandana and Kunwar Jalees. 2009. *Why is Every 4th Indian Hungry? The Causes and Cures for Food Insecurity.* New Delhi: Navdanya.

Silvern, Steven. 2002. "State Centrism, the Equal-footing Doctrine, and the Historical-legal Geographies of American Indian Treaty Rights." *Historical Geography* 30: 33–58.

Slocum, Rachel. 2006. "Anti-racist Practice and the Work of Community Food Organizations." *Antipode* 38(2): 327–349.

Slocum, Rachel. 2007. "Whiteness, Space, and Alternative Food Practice." *Geoforum* 38(3): 520–533.

Slocum, Rachel. 2008. "Thinking Race Through Corporeal Feminist Theory: Divisions and Intimacies at the Minneapolis Farmers Market." *Social and Cultural Geography* 9(8): 849–869.

Smith, Fiona, Helen Timbrell, Mike Woolvin, Stuart Muirhead, and Nick Fyfe. 2010. "Enlivened Geographies of Volunteering: Situated, Embodied, and Emotional Practices of Voluntary Action." *Scottish Geographical Journal* 126(4): 258–274.

Smith, Neil. 1992. "Contours of a Spatialized Politics: Homeless Vehicles and the Production of Geographical Scale." *Social Text* 33: 54–81.

Spencer, James. 2004. "People, Places, and Policy: A Politically Relevant Framework for Efforts to Rescue Concentrated Poverty." *The Policy Studies Journal* 32(4): 545–568.

Stark, Stacey, David Abazs, David Syring, Gayle Nikolai, and Mike Mageau. 2010. "Defining the Agricultural Landscape of the Western Lake Superior Region: Realities and Potentials for a Healthy Local Food System for Healthy People." Minneapolis: Healthy Foods Healthy Lives.

Tarasuk, Valerie and Joan Eakin. 2005. "Food Assistance Through 'Surplus' Food: Insight from an Ethnographic Study of Food Bank Work." *Agriculture and Human Values* 22: 177–186.

Taylor, Alan. 2002. *American Colonies: The Settling of North America, Volume 1.* New York: Penguin Books.

Thrift, Nigel. 2007. *Non-representational Theory. Space, Politics, Affect.* New York: Routledge.

Tönnies, Ferdinand. 1957 [1912]. *Gemeinschaft und Gesellschaft.* Translated by Charles Price Loomis as *Community and Society*, East Lansing: Michigan State University Press.

Trudeau, Dan. 2008. "Junior Partner or Empowered Community? The Role of Non-profit Social Service Providers Amidst State Restructuring in the US." *Urban Studies* 45(13): 2805–2827.

UN Food Programme. 2010. Fighting Hunger Worldwide: The World Food Programme's Year in Review. Rome, Italy.

US Fish and Wildlife Service / US Census Bureau. 2007. 2006 National Survey of Fishing, Hunting, and Wildlife-associated Recreation. Washington DC.

Vileisis, Ann. 2009. *Kitchen Literacy: How We Lost Knowledge of Where Food Comes From and Why We Need to Get it Back.* Washington DC: Island Press.

Wacquant, Loïc. 2009a. *Punishing the Poor: The Neoliberal Government of Social Insecurity.* Durham: Duke University Press.

Wacquant, Loïc. 2009b. *Prisons of Poverty.* Minneapolis: University of Minnesota Press.

Walton, John and David Seddon. 1994. *Free Markets and Food Riots: The Politics of Global Adjustment.* Oxford: Blackwell.

Watts, Michael and Hans Bohle. 1993. "The Space of Vulnerability: The Causal Structure of Hunger and Famine." *Progress in Human Geography* 17: 43–67.

Weis, Tony. 2007. *The Global Food Economy: The Battle for the Future of Farming.* London: Zed Books.

Whiting, Erin and Carol Ward. 2010. "Food Provisioning Strategies, Food Insecurity, and Stress in an Economically Vulnerable Community: The Northern Cheyenne Case." *Agriculture and Human Values* 27: 489–504.

WHO (World Health Organization). 2004. "Commission on Social Determinants of Health." EB115/35. Geneva: World Health Organization.

Williams, Andrew, Paul Cloke, and Samuel Thomas. 2012. "Co-constituting Neoliberalism: Faith-based Organisations, Co-option, and Resistance in the UK." *Environment and Planning A* 44: 1479–1501.

Wilson, Amanda. 2013. "Beyond Alternative: Exploring the Potential for Autonomous Food Spaces." *Antipode* 45(3): 1719–737.

Wilson, William Julius. 1996. *When Work Disappears: The World of the New Urban Poor.* New York: Vintage Press.

Winne, Mark. 2008. *Closing the Food Gap: Resetting the Table in the Land of Plenty.* Boston: Beacon Press.

Wolch, Jennifer. 1990. *The Shadow State: Government and Voluntary Sector in Transition*. New York: The Foundation Center.

Wolff, Eric. 1982. *Europe and the People Without History*. Berkeley: University of California Press.

Wood, Dolores, Jill Armstrong, Miriam Edlefsen, and Sue Butkus. 2008. "Food Coping Strategies Used by Food Pantry Clients at Different Levels of Household Food Security Status." *Journal of Hunger and Environmental Nutrition* 1(3): 45–68.

Wrigley Neil, Daniel Warm, and Barrie Margetts. 2003. "Deprivation, Diet, and Food Retail Access: Findings From the Leeds 'Food Deserts' Study." *Environment and Planning A* 35: 151–88.

Yapa, Lakshman. 1996. "What Causes Poverty? A Postmodern View." *The Annals of the Association of American Geographers* 86(1): 707–728.

Young, Iris Marion. 1990. *Justice and the Politics of Difference*. Princeton: Princeton University Press.

Index

accompanying the poor 139
Alexander, Michelle 39, 47, 56, 113, 136
Alinsky, Saul 14, 50, 91, 138, 146, 147, 152–3
allostatic load 79, 106
alterity 3, 24, 100
anti-community 150

care economy 18
charity 8, 14–15, 19, 20, 26, 36, 45, 56, 62, 64, 107, 111, 112, 120–1, 134, 136, 138–9, 145, 148, 150
cheap food 65, 138
Christianity 9, 15, 26, 58, 61, 67, 83, 87, 89, 119, 141
CHUM Food Shelf (CHUM) 9–10
Cloward, Richard 21, 30, 32, 36, 38, 113
commoditization 23
commodity: and crops 30, 34–5, 112, 144, 147; and foods 3; and government 10, 12, 34–5, 149
community gardens 1, 73, 75, 81, 104, 141–4
community organizing 10, 14, 15, 23, 24, 32, 37, 46, 48, 51, 63, 82–5, 145, 147, 151

Day, Dorothy 112
Debord, Guy 76–7
defetishization 77
diabetes 31, 33

emerging projects 21, 44
employment 21, 36–37, 52, 112–14, 124, 126; and youth 141; and transitional 12–14, 95–8

faith 19, 24, 26, 42, 53–4, 58–63, 147, 148–9; -based organizations 82–3, 109, 119, 133
Feeding America 2, 28, 42
FNB vii, 45–6, 149
food desert 11, 12, 32–4, 84, 93–4, 151
food quality 11, 12, 14, 75, 88, 129
food quantity 117
food stamps *see* Supplemental Nutritional Access Program (SNAP)
Foucault, Michel 40–1

Gibson-Graham, J.K. 3, 21, 43, 88
governmentality 40–1, 113
grassroots 40, 55, 63, 138, 153

health insurance 126
Heynan, Nik 4, 18, 37, 45, 81, 145, 149
homeless 1, 9, 21, 30, 45, 54, 72, 76, 79, 82, 84–6, 111, 126–8, 132–3

Kropotkin, Peter 45, 77, 111

liberation theology 37, 139, 142, 149,
liminal 4, 47, 112, 122, 153

Malthus 31–3
marginalization 3, 4, 18–19, 22, 26, 29, 31–2, 36, 46, 48, 56, 96, 136, 143
Maurin, Peter 77, 80
mental health 17

neoliberalism: actually existing 20, 95; defined 19–21; pious 14–15, 119–20; roll-out 17, 20, 83, 135

168 Index

Piven, Frances Fox 21, 30, 32, 36, 38, 113
politicization 144–8
poor relief 2, 30, 34, 38, 47, 138, 147
Poppendieck, Janet 2, 17, 26, 28, 34, 46, 49, 82, 135, 138, 150
prison 20, 39, 97, 113, 121, 136

Ruby's Food Pantry (RFP) 9

scale: and jumping 84; and local 42–3; and program size 50, 76–7, 79, 83, 109, 138, 150
Seeds of Success (SoS) 10–11
Self-help and Resource Exchange (SHARE) 8–9
spaces of hope 105, 137
stigma 2, 14, 19, 22, 30, 46, 62, 99, 101, 103, 107, 109, 113, 115–17, 119–20, 128–9, 132

Supplemental Nutritional Access Program (SNAP) 3, 144, 147, 152
surplus 2, 9–10, 15, 25–6, 28, 31–2, 34–6, 43–4, 76–7, 80–2, 87–91, 134, 142, 145
sustainable compassion 59–62

unemployment 6, 80, 86
USDA viii, 18, 33, 39, 97, 114–15, 144, 146–7

volunteerism 119–20,

whiteness 57, 67, 77
WIC 3, 13, 25, 34–5, 69, 114, 120, 126, 128, 130–1, 144, 147
workfarism 37–42, 94, 112, 133–4, 137, 141–2, 148

For Product Safety Concerns and Information please contact our
EU representative GPSR@taylorandfrancis.com Taylor & Francis
Verlag GmbH, Kaufingerstraße 24, 80331 München, Germany